Simulation Training: Fundamentals and Applications

Adapted from a work first published in French under the title: "*Améliorer les pratiques professionnelles par la simulation*" (© Octares 2011)

Philippe Fauquet-Alekhine
Nane Pehuet

Editors

Simulation Training: Fundamentals and Applications

Improving Professional Practice Through Simulation Training

Forewords by Prof. René Amalberti and Prof. Jacques Leplat

 Springer

Editors
Philippe Fauquet-Alekhine
Chinon Nuclear Power Plant
Avoine
France

Nane Pehuet
CNRS-INSHS
Paris
France

ISBN 978-3-319-19913-9   ISBN 978-3-319-19914-6   (eBook)
DOI 10.1007/978-3-319-19914-6

Library of Congress Control Number: 2015942217

Springer Cham Heidelberg New York Dordrecht London
© Springer International Publishing Switzerland 2016

Printed on acid-free paper

Springer International Publishing AG Switzerland is part of Springer Science+Business Media
(www.springer.com)

Foreword I

Thanks to the authors for having called me to write this foreword. I had even more pleasure to read than to write.

Perhaps I was not the more qualified for this honor as I was neither a professional of the simulation nor of training. I thus had to recall my memory and my remembrances accumulated here and there in transport companies, industries, and medicine, yet the effort has not been so heavy as I often have been kept close to training as a natural continuation of the human factors and organizational interventions; I thus saw with time the power of the concept "simulation training" increasing, as a magic idea, essential, impossible to be passed around.

I accompanied the beginning of the concept in the 1980s and the early 1990s with merchants challenging to catch the new market, asking more real than real, finally selling simulators awfully heavy to use; I have seen trainers who had never been involved in the real work activity and I have seen operators, broken by illness or career errors, positioned as a trainer almost as a punishment; I have seen these failures and I have also seen lessons rapidly learned by all the stakeholders, industrials as well as trainers. The result of all that was the birth of a knowledge core devoted to the use of these training tools, the birth of the academic side of this field, of know-how, of all that what this book tells us so well about.

Why such a success? ... Perhaps because simulation remains the last tool allowing the meeting of an inaccessible real. Thinking about that I identified at least four reasons which seem to have imposed this ineluctable rise:

- The safety development of industrial plants and risky professions has progressed quickly. As a paradox, this had a double negative effect for training: impoverishment of the contact with the risk and rejection of this risk. The real operating situations, monitored as a routine, do not offer anymore the training field of the past. Even more, training for manual operating, without any protection, is not anymore wished; it becomes difficult to get out of the procedures and assisted modes, both for safety questions and moreover for performance improvement purpose. Briefly, simulation offers the access to a real that reality does not offer anymore.

- Companionship has been shattered everywhere; the seniority has lost of its aura for the benefits of academic diploma and qualifications; official time length for training and official training programs are today more official, stamped on professional books, absolute sesame of the professional progress almost independent of the real value in the job; these counted times have replaced the informal hours spent to listen to and to watch experienced workers in the past trades. For all these reasons, training appears farther from the work activity, but has found back in the simulation training the natural trait that could have missed.
- The multiplicity of technological artifacts has also favored the simulation training. Helps, especially through computer science and electronics, induce a great complexity of the architecture of surfing that must be blank tested. Moreover, the technical specificity of these artifacts fits perfectly the simulation as they create by themselves layers that can be isolated (disconnected from the process environment) like training objects.
- Finally, and not the least, the know-how of the simulation has become available, every day more sophisticated, to better simulate the complexity of the real.

Not surprising in these conditions that the training of high-risk professions has been forced to go this way.

It just has to solve the common equations of any training:

Simulating: why (for which benefits)?

- The answer seems simple: the training, as addressed by this book (even if the conception or research usages are both pertinent). But the detailed answer about the "why" is less trivial than it seems to be. The initial choices prepare the final success of the pedagogical edifice; nonetheless, they sometimes are quite divergent regarding the scenario to be used. One can count several main inputs, all pertinent and all different: training to the procedural operating, to the general principals, to the abnormal situations (listed), or even to exceptional situations (by their very nature), etc.

Simulating: what?

- The technical system in its whole reality more or less simplified; it is obviously what we think about first…
- But we must not forget the scenario, reflecting the reality thought or elaborated according to some observed accidents/incidents for which physical actions (defects, incidents), human actions (errors), and added systemic actions (lightning, a unexpected director, the economical context); everyone will conceive that the degrees of freedom of the technical simulation, already numerous, are nothing compared to the infinite possibilities to elaborate scenarios. Anyway, the key of the successful training is the elaboration of the scenarios, not the fidelity of the simulation.

- Beyond the technical constrains and those of the scenarios, the simulation sessions are also a request to learn in a limited time and according to a constrained program. Paradoxically, this constriction of time and events, this acceleration of the real to make it expressing a set of pedagogical situations imposed during the time of the training exercise, is the main deviation to the real without any doubt and carries on some impressive difficulties for elaborating the scenarios.

Simulating: how?

- One of the medical chapters of the book evokes the pig as the first surgery simulator; we can see the infinite possibilities available for the trainers, from the simple alive cousin to the tool of cognitive equivalence assumed to involve the same intellectual qualities (the micro-worlds), to the realistic simulator (trustful on the surface, which recreates the climate, but whose equations of processes are greatly simplified in depth), or to the most faithful possible complete simulator (but is there actual such a simulator when the copy has necessarily implied a choice of arbitrary framing of the real, never reflecting all of the connections that keeps on this technical sub-object with its neighbors, with its plant and its environment).

Simulating up to which point of loss of control for a successful training?

- Professional simulation is not a game. It gives to see skills and failures, questioning the ego of the professionals and can generate real emotion of the participants. These dimensions can usefully serve the elaboration of skills, but their gradation must be particularly controlled and probably applied with judgement according to the operators' personality, because trainee is the target of the training installation, and trainee is also the only piece of the puzzle that is not simulated therefore to be handled with the greatest caution.

Beyond these issues, some recurrent questions should not be hidden regarding these simulations for training: which cost (the simulator, its purchase and maintenance, but much more the immobilization of the professionals), for which actual benefits (with questions on assessment), with which trainers and which training for the trainers?

All these questions (and more) find answers in this book.

It is a practical book about the unvarnished simulation, without too many scientific jargon, but written by those who do, and who already faced failures, in very varied applied areas, as metallurgy, nuclear industry, medicine, and transportation. All of them are true professionals in training on simulators; furthermore, they took time to think about their own practices, having digested the scientific knowledge.

It is also a book on training beyond a book on simulation, which is natural since the tool is a necessary condition but is never sufficient to achieve the objective...

I had great pleasure to read it, and I found that it was finally a rare reading by its simplicity, essential for countless candidates aiming to the use of simulators for training.

Good reading!

René Amalberti

Doctor in medicine and psychology, René Amalberti was military physician, chargé de mission Human Factors for the Directorate General of Civil Aviation (DGAC), associate researcher at the French National Research Center (CNRS), Deputy Director of the Institute of aeronautics medicine, Professor at Val de Grace Hospital in Paris (France), and researcher at the IMASSA-CERMA. His main areas of research are cognitive ergonomics, cognitive modeling of the operators. He is the author of numerous books on high-risk systems making references now on the international scene. He now divides his time between the High Authority of Health (HAS) and medical insurance (MACSF).

Foreword II

The title of the book states clearly the spirit of its content and how it should be read: It was written by trainers for trainers. The authors do not seek to engage in a general debate on the simulation, but to identify under which conditions it can be used to improve professional practices. In this aim, they decided to reflect together on their respective practices engaged in various sectors: airline, medicine, nuclear industry, and metallurgy. Despite the variety of studied situations, it quickly appeared that these exchanges showed many common traits to the approaches taken by each contributor and that each one could gain from a better understanding of the others. Wherefrom the idea of extending the reflections in developing a book which could be enjoyed by the trainers' community, not only in seeking to synthesize the drawn principles, but also in developing very concretely and specifically the analysis of cases from their practice. Thus, the importance appears of an in-depth analysis of the trainees' activity. The conceptual chapter will help readers to develop their own synthesis on the simulation method and improve their personal practice.

This book, well organized and well presented, was designed primarily for trainers, but it is likely to be appreciated by all those who have to deal with occupational training.

Jacques Leplat

Jacques Leplat, nowadays Directeur Honoraire at the Ecole Pratique des Hautes Etudes (EPHE, Paris, France), has been managing for a long time the Laboratory of Work Psychology at EPHE and has been co-manager of the revue "*Le Travail Humain*" (human work). He is the author of numerous publications in the field of cognitive and ergonomic psychology, especially addressing the analysis of the individual and collective activity in work situations regarding training, safety, and reliability. He has received the "Ergonomics Development Award for outstanding contributions to the development of ergonomics internationally" from the "International Ergonomics Association" (1994).

Preface

In 2005, the French research association ECRIN[1] suggested the book entitled *"Apprendre par la simulation"* (Learning through simulation), resulting from the long collaborative work between researchers and practitioners, directed by Professor Pierre Pastré. From the analysis of learning activities, it was suggested to the reader a reflection about the conception of simulation situations improving learning. This first book, situated within the occupational didactic theory, has shown the broad diversity of professional training through simulation, and has permitted to point out the importance of methodological questions linked to training practices with these new tools, such as re-contextualization of the task, the evaluation, and the debriefing...

In 2010, in view of these considerations both theoretical and practical, it appeared useful to extend the reflection by investigating the most actual usages of simulation within different professional environment.

The present book is thus proposed as the following of the first one, centered on the training method rather than on the learning situation. Furthermore, we have written from the standpoint of the trainer, the engineer, or the trainee, so that methods related to simulation training practices could be presented in a concrete manner and resorts could be emphasized. Everyone will be able to use it according to one's own expectations and needs: to enhance the management of simulation situations or to improve one's professional practice, to optimize one's reliability or performance, for maintenance or operating activities ...

For each described situation, the reader will be given both pragmatic advices for the practice and theoretical complements on which rely these propositions.

Philippe Fauquet-Alekhine and Nane Pehuet, both at the origin of this project, have constituted a work group in the frame of ECRIN association, gathering experts of the simulation training practicing in a broad range of professional activities such

---

[1]Exchange and Coordination Research-Industry, ECRIN was created in 1900 by the French national research center (CNRS) in the aim to favor exchanges between public research laboratories and industries and to promote innovations.

as aircraft pilots, anesthetists, surgeons, metal-rolling mill operators, and nuclear reactor pilots. All these experts, from the operating or training world, had all in common a trainer's experience. Each of them has taken in charge one chapter according to the same frame: the professional context description, the pedagogical methods, weaknesses and strengths, teaching formalism, and pragmatic advices.

In the book, authors present their own experience and give evidence of their own constraints regarding their profession, the organization, or the company in which they work. We wish readers will find in this book solutions that will permit to solve some difficulties encountered, and some advices useful to make their own professional practice to evolve regarding simulation training methods and more specifically the articulation between the simulator and the simulated situation.

There are several usages of the simulation for occupational training. We have chosen to investigate initial training (anesthetists, surgeons) as well as training designed for experienced professionals (rolling mill operators, aircraft pilots, reactor pilots) for whom the simulation training is a mean to maintain, to make reliable, or to improve their professional practices. It has become fundamental for the companies to maintain competencies and skills of the teams, at least for security, safety, or reliability reasons.

We would like to highlight now that for each training context, a specific method exists which fidelity degree compared to the professional reference situation is variable. Some simulators presented thereafter are said "full scale" or "full scope" in the sense they fulfill a complete transposition of the real operating facilities, while others, rather targeting specific learning (professional gesture for example), will be limited to a partial transposition of the real operating situation.

The first chapter suggests a general reflection shedding light on the pedagogical methods applied through the five following chapters. It aims to clarify and to emphasize the theoretical hypotheses which underlie pedagogical practices through selected examples in the following chapters.

The next chapters are ordered with regard to the type of simulator use for training: The two first ones concern the aircraft pilots and the nuclear reactor pilots, two activities that require full-scale simulators for which the likeness is remarkable. Then, a chapter comes addressing anesthetist training. Here, again we have a full-scale simulator, but with a major difference regarding the relationship between the simulator and the trainees: Anesthetist training takes place "around" the simulator, while for the two first cases, training takes place inside the simulator. And finally, the last two chapters present a hybrid simulator for the surgeons' training and rolling mill simulator. The hybrid simulator for the surgeons' training combines virtual reality and situation close to the reality, with surgeons trained with the help of imagery software over organs from which they can feel the "matter" due to feedback effort systems. The rolling mill simulator is a kind of full-scale simulator tending to a virtual simulation.

All along the book, inserts have been added to specify technical points and some lines are highlighted to emphasize pragmatic advices. A thematic table is put at the end of the book and helps the reader to find out easily the main items.

With this book adapted in 2015 from the French 2011 version and updated with recent bibliographic references, we wish the reader to feel the great possibilities offered by the simulation tools toward occupational learning and training, keeping in mind that the tool alone is not sufficient and must be necessarily inscribed inside a structured and coherent pedagogical whole.

Philippe Fauquet-Alekhine
Nane Pehuet

Acknowledgment

This book has been realized in the frame of a work group constituted under responsibility of the ECRIN association and supported by the Laboratory for Research in Science of Energy. Author and contributors are hereafter presented one by one. Special thanks to Professor Jacques Leplat for his contribution to the work group.

Contents

Editors and Contributors

About the Editors

Nane Pehuet As CNRS engineer, project manager until 2008 at the Club "Developments of work facing technological mutations," she has coordinated several collective works such as "Apprendre par la simulation" (Learn from the simulation) under the direction of P. Pastré and Ed. Octares (2005); "Facteurs Humains et Fiabilité" (Human factors and reliability) under the direction of R. Amalberti and Ed. Octares (1997); and "Travail et competence, le sens des evolutions" (Work and skills: the meaning of developments), Ed. Club ECRIN CNRS (1998).

Since 2009, she was responsible for the valuation of the national institute of Human and Social Science of the French National Research Center (INSHS/CNRS) and is now in charge of European and international cooperation.

She is graduated from the University of Paris-Dauphine in Sociology, graduated in Science of the Education from the University of Paris 1 Sorbonne, and graduated in science andtechnology from the Conservatoire des Arts & Métiers (Paris, France). e-mail: nane.pehuet@cnrs-dir.fr

Philippe Fauquet-Alekhine Human Factors Consultant, doctor in Physics Science, Work Psychologist from the Conservatoire National des Arts & Métiers (Paris, France), author of several scientific communications, he contributes to researches and interventions in firms regarding the study of Human in work situation, work organization, and management; Member of the Laboratory for Research in Science of Energy, he collaborates to research in psycho-sociology at the Department of Social Psychology (LSE, London, UK) and at the Medical Training Center of the University Hospital of Angers (France). His scientific productions especially concern the analysis of work activity, its modalities, contributions and its application in industrial environment. They also concern more specific sides as the psycho-linguistic approach for the analysis or operational communication, or cognitive aspects of non-simulated work activities or of learning and training on simulator. For the industrial field, he investigates aerospace, airlines, air force army, police, navy, nuclear industry, and medicine.

Involved in the pedagogical conception of experimental training on simulator, he co-elaborates new scenario based on his own experimental observations or by taking advantage of his experience as safety expert on nuclear power plant (NPP) during four years, trained as reactor pilot in accidental situations and involved in exercises of crisis management. NPP of Chinon, BP80, F37420 Avoine, France. +33 (0) 2 47 98 7804 e-mail: philippe.fauquet-alekhine@edf.fr, http://www.hayka-kultura.org/

Contributors

Gérard Bonavia Until 2011, Gérard Bonavia was responsible for the formalization and the transfer of professional knowledge at Arcelor Mittal Fos-sur-Mer plant in the context of a very strong generational renewal. Having started his career in the steel industry in 1974 as store electrician working in 3×8 and then having held positions in the electrical maintenance of the company, he moved toward the trades of computer maintenance. In 1983, he joined the technical service of a hot rolling production team to work on the improvement of manufacturing processes. He began studies on modeling and implementation of automation. He then analyzed the role of operators in the process and how to help training for their profession. In 2001, he designed a project for the development of skills of manufacturing. Stéphanie Guibert integrated the project group, with computer scientists and operators enforcing the team. This team imagined and built the training simulator and the pedagogical support system. This simulator is used today by any other rolling mills in the north of the France and Spain. In 2007, G. Bonavia conducted a project to build a training strategy, and in 2008, he joined the training department to implement it. Since 2012, G. Bonavia went back to the engineering of steel industry as private consultant. e-mail: g.bonavia@free.fr

Fabien Trabold Anaesthesiologist Resuscitator and Commander, Chief Medical Officer. By his double activity as firefighter commander and doctor in anesthesiology and intensive care, he faces on a daily basis the management and organization of care in crisis situations. He is certificated in disaster and emergency medicine. He is the initiator for many years of the development of the simulation in the medical field for acts in emergency.

Haut-Rhin Fire Department Rescue and Health Service (SDIS 68), 68027 Colmar Cedex, France; Department of Anesthesiology and Intensive Care Unit, Bicêtre Hospital, 94275 Kremlin Bicêtre, France. e-mail: fabien.trabold@sdis68.fr

Thomas Geeraerts Anesthetist Resuscitator, Anesthesiologist and Intensive Care physician Clinician and researcher, Thomas Geeraerts is Professor at the University of Toulouse and specialist in anesthesia and intensive care; he holds University neuroscience PhD. His main clinical activity is in Neuroanesthesia and

Neurointensive Care. He was part of the simulation group at the Hospital of Bicetre (public hospitals of Paris) since his inception and has contributed to multiple courses for students in medicine, residents, and doctors who specialize in anesthesia and intensive care. Since November 2009, he took his duties in Toulouse hospital, where he participated in education on critical situations in anesthesia. He created the simulation center in Toulouse in 2014. Department of Anaesthesia and Intensive Care, University Hospital of Toulouse, Toulouse, France. + 33 (0) 5 61 77 21 45, e-mail: thgeeraerts@hotmail.com

Marc Labrucherie is a consultant in the field of Human Factors and the reliability of the operators at risk in complex systems (NML Consulting).

Former student of National Civil Aviation School (ENAC), he has exercised the profession of aircraft pilot at Air France from 1973 to 2008 and, at the end of career, has 21,000 flying hours including several thousand as instructor and controller. He has also held several executive positions including head of the Training Department on Boeing 747.

Moreover, graduated in physiology from UTC (University of Technology of Compiègne, France), he created and animated the Human Factors group in Air France in 1985, and implemented Crew Resources Management training, and then in 1993, he created the Human Factors Department of the company.

Co-author of the audiovisual course DEDALE-IFSA "Briefings", he also established the bank of questions and worked during 6 years, on behalf of the Board of Examinations of the DGAC (Direction Générale de l'Aviation Civile), as manager of the human factors exams for certification of pilots of aircraft and helicopters.

Furthermore, he participated in many conferences for various organizations including the ICAO (International Civil Aviation Organization). e-mail: malabrucherie@gmail.com

Jacques Marescaux President and Founder of the IRCAD Institute, Chairman of the Institute of Image-Guided Surgery received his doctorate in medicine at the University of Strasbourg in 1977. In 1982, he joined a team of INSERM and became professor at the Digestive Surgery department. He created the Research Institute against Cancers of the Digestive Apparatus (IRCAD) and the European Institute of Telesurgery (EITS) in 1994. Prof. Marescaux is a member of a large number of international medical societies. He is recognized internationally for the first long-distance surgery between New York and Strasbourg, and for this performance he was nominated for the Computer World Honors Award in 2003 from the renowned Smithsonian Institute. He also received a very large number of national and international awards for his work in the areas of computer-assisted surgery, surgical simulation, surgery, surgical robotics, and more recently transluminal surgery. University of Strasbourg, Strasbourg, France; IRCAD, 1, place de l'Hôpital, 67091 Strasbourg Cedex, France. +33 (0) 3 88 11 90 06, e-mail: jacques.marescaux@ircad.fr

Carole Maridonneau was involved as trainer in simulation situations for nuclear reactor pilots at the Training Center of Chinon, France, working on the practice analysis for the operating teams, contributing to the elaboration of new scenario referring to reference situations, and applying the method of exploitation according to the school of Vermersch during the debriefing.

She was then working as Human Factors Consultant at the new generation nuclear power plant built in Flamanville (France): the European Pressurized Reactor (EPR) delivering 1600 MWe and at the NPP of Chinon. e-mail: carole.maridonneau@edf.fr; c.saphore@orange.fr

Luc Soler is R&D Director in computer sciences within the Research Institute against Digestive Cancer (IRCAD) in Strasbourg, France. In October 2000, he also joined the digestive and endocrine surgery team of Professor Marescaux as invited professor at the Medical Faculty of Strasbourg. In July 2013, he furthermore became president of the Visible Patient S.A.S. company. His main areas of interest are medical image processing, 3D modeling, virtual and augmented reality, surgical robotics, and abdominal anatomy. In 1999, his research work has been awarded with a Computer World Smithsonian Award, in 2003 with the first World Summit Award in the eHealth category, in 2004 with the "Best Vision Paper" of IEEE Robotics & automation society, in 2005 with the 2nd international award of the "Sensable Developper Challenge," in 2006 with the "Le monde de l'informatique" trophy in the Health category, and in 2008 and 2009 with the first prize of the Kitware Best biomedical Vizualization award of MICCAI. e-mail: luc.soler@ircad.fr

Christine Vidal-Gomel Ph.D. in psychology and ergonomics, lecturer in the Department of Science of Education in the University of Nantes (France), her research topics are as follows: development of professional skills, training design, and the activity of trainers [recent publications in the reviews Work: A Journal of Prevention, Assessment and Rehabilitation and Applied Ergonomics, and in the collective books Constructive Ergonomics edited by P. Falzon (published by Taylor & Francis), Risk and Cognition edited by J.-M. Mercantini and C. Faucher (published Springer)]. e-mail: christine.vidal-gomel@univ-nantes.fr

Reflections and Theoretical Contributions Regarding Trainers' Practice and Simulation

Christine Vidal-Gomel and Philippe Fauquet-Alekhine

It is not surprising that a number of trainers met in companies ask themselves recurring questions about what they have to do, if they do it well, if they could do it better... Testimonials in following chapters will provide elements for reflection and will try to provide answers. This chapter proposes to bring to light some conceptual aspects related. We hope to demonstrate that theoretical inputs will help to better understand the term "practice" and help "to practice better". In other words, if we consider "the practice as an object of analysis", as said by Marcel et al. (2002), our point of view on the "practice", which is a rather approximate term, is to promote understanding of the professional act, for implementing this debate and possibly make it evolving, from theoretical inputs produced by ergonomics, ergonomic psychology and two currents of thought which have been inspired: professional didactics and the clinic of the activity, and from what the workers themselves tell us about it. Everything will not be said, but we hope to take crucial points and to give elements of understanding.

In the following chapters, the authors will describe different possibilities of operation of the training by the simulation. These include: initial training, development, training in the application of special procedures such as those that allow to deal with potentially risky situations, or the work of rare situations. Let us remind that initial training is not addressed in the present book. In the simulations we are dealing with, the bases of the trade are already integrated by the trainees, and in most cases, they become the support of the training.

To take this into account, the first part of this chapter will be devoted to the development and management of simulations by trainers. We shall then look at the

C. Vidal-Gomel
Department of Science of Education, University of Nantes, Nantes, France
e-mail: christine.vidal-gomel@univ-nantes.fr

Ph. Fauquet-Alekhine (✉)
Laboratory for Research in Science of Energy, Montagret, France
e-mail: philippe.fauquet-alekhine@edf.fr
URL: http://www.hayka-kultura.org

Ph. Fauquet-Alekhine
NPP of Chinon, BP80, 37420 Avoine, France

© Springer International Publishing Switzerland 2016
Ph. Fauquet-Alekhine and N. Pehuet (eds.), *Simulation Training: Fundamentals and Applications*, DOI 10.1007/978-3-319-19914-6_1

way in which the people involved in the simulations are brought to build compromises between conflicting objectives. For trainers, it can be educational goals opposed to the cost of a simulation or the number of these trainers necessary for the operation of the simulation. To conclude this chapter, we shall discuss the training of trainers, who have more often an "operational" experience, and the evolution of their skills from industrial workers to trainers.

1 Developing and Using Simulation Situations as a Training Objective

Developing and using simulation situations as a training objective requires several types of questions that can be listed from a framework of analysis proposed by Samurçay (2005). We will use a schematic and simplified representation (Fig. 1).

The framework proposed by Samurçay (Fig. 1) considers three sets of questions enclosed together and articulated, regarding respectively (i) the simulator, i.e. the artifact,[1] (ii) the simulated situation, that can be set in the first place as the specific situation which was designed for training and (iii) the simulation situation, i.e. the situation simulated as it evolves because of the activity of the trainees and the activity of the trainers.

These three sets of issues will be used here as a frame of reference for theoretical inputs in relation to the issues discussed in following chapters.

1.1 Simulators

In the proposed framework, the simulators are defined as the artifacts that simulate (partially or completely) the operation or the behavior of a technical system, facility, or a natural phenomenon. Their design requires models (Fig. 1) of the simulated device, and possibly of the external environment. These devices are equipped with interfaces for the interns and optionally interface to the trainer(s). The characteristics of the simulator are an important determinant of choice which may or may not be made about the simulated situation.

Thus, taking account trainers as soon as possible in the design of the artifact can be a real issue: this is to allow them to contribute to the reflection on the possibilities of simulating some types of situations that they could be working by the forms, to give access to the information they need to understand the activity of the

---

[1]In the common language, we could say "the tool". To be more precise, in the frame of activities with instrument defined by Rabardel (1995), an artifact (physical or symbolic) is elaborated through the human activity or transformed by it and is possibly used, that is being a part of a finalized activity.

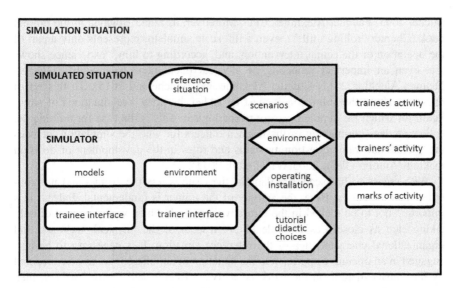

Fig. 1 Three levels of analysis of simulations adapted from Samurçay (in Vidal-Gomel 2005)

trainees in situation, in order to act on didactic variables of the situation (having an effect on the complexity of the situation on the basis of skills that they have identified, for example) and to organize the phases of debriefing and promote the analysis of activity afterwards.

Simulators to which we are interested in the present chapter are "full scale simulators" (or "full scope simulator"), in that they reproduce in its entirety (with almost all details), the human-machine interface which is the one of the real work of the operators, both technically and in terms of perception (the worker sees, hears, touches an industrial environment identical to that of the real operating situation). Thus, the full scale simulator combines operational and figurative aspects (within the meaning of Pastré 2005), the closer to the reality of operation so that we might say that the trainees perceive themselves "in action" in this "reality of operation", even if the situation is simulated.

These simulators which have a high degree of fidelity are also very expensive, both in the design or purchase, and in the use, as remind Soler or Geeraerst and Trabold (Chapters "surgery" or "anesthesia"). Also, the question is asked regarding the use of partial simulators or simulators of "problem solving", as named by Laborde and Pastré (2005), and concerning their interest in training. For example, Nyssen (2005) compares a simulator using full scale and a partial simulator in anesthetists' training. She thus shows learning does not vary significantly, although the pace is different. Patrick (1992) had already noted that the fidelity of the simulation is not necessarily a guarantee of learning. Other works highlight "transformation of the real", operations that enable new technologies such as virtual reality to be useful to learning and vocational training (Mellet d'Huart and

Michel 2005; Fauquet-Alekhine 2011). Moreover, as noted Bonavia in the present book (Chapter "rolling mills"), even a full scale simulator represents only a part of the operation or the actual intervention, and, according to him, "experience shows that even an imperfect modeling of physical phenomena is enough" (see also Fauquet-Alekhine 2011; Fauquet-Alekhine and Boucherand 2015). To this set of questions about the simulator, it must be added that there are situations of simulation for which the simulator is not technological. This is the case for training the collective decision-making of the firemen officers for who the simulation includes the operational frame questioning status and roles in the development of decision making (Antolin-Glenn 2005; Vidal et al. 2011).

With regard to full scale simulations, the issue is less the artifact and its properties than their uses. Here, the activity of the trainer is fundamental. Indeed, it is important not to be locked in the figurative approach which would consist of only taking care as closely as possible of the respect of the physical, technical and organizational characteristics of the reference situation. It is necessary to be also engaged in an operative approach to the situation of simulation, i.e. to be concerned by the impact thereof on the workers' activity (Béguin and Pastré 2002). It then means to take into account, at the same time, the activity to be theirs in a simulated situation, the activity which will be theirs in a later industrial operating situation, and to make the simulated situation a mediator for the activity development in a real operating situation. In other words, in training, the question is therefore that of mediation: mediation by the trainer, by the simulator artifact, and also by the simulated situation (Samurçay and Rogalski 1998).

1.2 The Simulated Situation

The simulated situation can be defined as the product of the didactic transposition of the reference situation (Samurçay and Rogalski 1998; Samurçay 2005). The reference situation represents the category of situations which trainees must be able to manage at the end of the training. It can undergo many types of transformations according to the training needs. These transformations characterize the transposition process. Several transformations of real situations can be operated for the purpose of training. Samurçay and Rogalski (1998) specify three: dividing (tasks and subtasks for example), decoupling variables [decoupling the technical dimensions from the security dimensions as for example can do training in the prevention of electrical hazards (Mayen and Vidal-Gomel 2005)], and focusing (it concentrates on relations between variables).

Other changes can be made to give access to the relevant information (Laborde and Pastré 2005). New technologies such as virtual reality, allow training designers to add information which are not accessible in a real situation in order to facilitate some aspects of learning (Mellet d'Huart and Michel 2005). In the case of a dynamic process to be monitored or managed, it is also possible to play on its tempo, by slowing down or by accelerating, so that it gives the trainees easier access to the results of their actions. For example, the simulator developed to promote the learning of the grapes cut allows the apprentice to work on the same vine plant three times in the same session, so that it matches the evolution of the plant for three consecutive years. The cut of the year "n" being taken into account in the growth of the grapes to year n + 1, etc. (Caens-Martin 2005).

The transposition in its different forms can meet progression in training objectives by integrating the need to not cognitively overload the trainees, so that they integrate more easily what they do on simulator. For example, aircraft pilots who are to be qualified on a new machine are not trained at first on simulator for landings with cross-wind and fog; they will be first in a situation to land on with favorable weather conditions. Conversely, the transposition may be to increase the complexity of the most common work situations to train for the management of rare cases, particularly complex.

To summarize, the transposition refers to all of the transformations that trainers choose relative to the reference situation, according to their own implicit "models" or explicit learning, and according to development of skills and the difficulty of what trainees should acquire.

The quality of transposition is crucial, because it partly determines the manner in which the trainees will be able to involve themselves "inside the situation", i.e. to act in a simulated situation that they will perceive as close to a operating situation or a real intervention, what the trainer is seeking. Soler (Chapter "surgery") speaks of "mental immersion in the performed surgical procedure". Geeraerts and Trabold (Chapter "anesthesia") illustrate how this mental immersion can be intense, by relating observations on simulator in which students live "intensity of stress [which] sometimes leads to surprising reactions (screaming, aggression, or complete inhibition)". If these events are not the first effects desired on simulator, we postulate that they nevertheless prove that the trainees "enter" in a situation and that they do all what they would be able to deploy in a non-simulated situation. In Hoc's terms (2001), these demonstrations attest to the "ecological validity" of the simulation.

Maintaining trainees "in situation" is fundamental because, to quote Pastré (2005), simulation situations must be "professional situations where an operator is obliged to transform his repertoire of skills to adapt it to a new professional condition." It is important for the trainer to bring attention to this phase of work.

Detailed examples are presented by Bonavia (Chapter "rolling mills") or Soler (Chapter "surgery").

The choices for the transposition will determine the possible scenarios (as the scenario of a film, they are a time slice of a fictional story in which the actors of the simulated situation should blend), the environment or the operational device (who does what? what roles should play the trainer in the simulation?). This implies for the trainer to manipulate two types of input data: technical parameters and socio-organizational and human parameters.

- The technical parameters are the setting of the simulator: the values of different variables involved in the scenario. For example, for aircraft pilots, these will be the load of fuel, wind speed, weather on the runway; for the anesthesiologist or surgeon, the age of the patients, their pathology, their allergies to the drug products, their history, amongst others.

- Socio-organizational and human parameters are those which structure interactions of the workers and the interfaces with the simulator, in other words the operational device (Fig. 1): who will occupy such function in the team? What information can the team get from another team supposed working beforehand on this situation? What is the support available in the organization at the time where the script will start? etc.

Other issues are to be considered as those of didactic and tutorial choices. One of the important points, for the high risk systems discussed in the present book, is to make individual as well as collective decisions during the simulation of "limit", "rare" and "risky" situations (see Chapter "reactors"). Labrucherie (Chapter "aircraft") for example asks the question of the pilot training in a 40-knot crosswind, knowing that they must avoid real flying situations in winds over 30 knots. This type of simulation helps to learn about the system and its limits.

With another point of view, Geeraerts and Trabold (Chapter "anesthesia") propose to improve the performance of the trainees facing a critical situation, even rare; and beyond appears the problem of leading scenarios up to unacceptable situations such as those resulting in the death of the patient, the airplane crash, the fusion of the reactor, or the loss of several tens of tones of steel. For example, for anesthetists, it was in the habit of denying the death of the patient in a simulated situation to not hit or hurt physicians for who, of course, the patient death is unacceptable. The trainers' approach is very conservative, as Geeraerts and Trabold write: the purpose is not to place the student in a counterproductive context on the educational standpoint or to hurt her/him. A study about the University training of anaesthetists (Fauquet and Frémaux 2008) in the anesthesia-reanimation Department of CHU Bicêtre in 2008 provides us an example. Prior to this training, meetings between researchers and physician-trainers had found acceptance to lead scenarios to the ultimate end, the patient death. However, during one of the simulated situations, a 3-trainee team (residents) was soon facing the death of the patient-manikin. The achievement of the unacceptable situation could have been effective: the residents appeared unable to find the vital functions necessary for the return of the patient. It was then that one of the physician-trainers made the decision to play the role of an onto-rhino-laryngologist doctor in reinforcement: they restored the patient's vital functions. The post-training discussion between researchers and physician-trainers found that they had made this recovery to save themselves from the consequences induced by the achievement of an unacceptable situation. Indeed, until then, the main question always concerned the manner in which the residents would accept the simulation result of such a situation, but never this had been raised for physician-trainers. Yet observations have shown that these physician-trainers needed to first develop a strong and coordinated strategy to cope with such a situation.

This development goes through the identification of possible result at the scenario end and of opportunities to work with this scenario, by asking for example how to terminate it or what to do if one of the residents bursts into tears.

Two lessons are to be drawn from this experience:

- reaching the unacceptable situation may be effective "without forcing scenario", leaving trainees implement their know-how and skills in the context that is proposed to them;
- the achievement of the unacceptable situation presupposes that a strong and coordinated strategy has been established beforehand by physician-trainers, both to preserve the residents and to ensure physician-trainers that they will be able to deal with the situation.

Labrucherie (Chapter "aircraft") explains that those who are advocates of not going to the crash of the aircraft are motivated by the desire not to "traumatize" the crew. However, he notices that it is also the opportunity for pilots to develop "creative survival strategies" rather than give up. For him, "it seems legitimate to assume that aviation professionals must be sufficiently strong on the mental aspect to face their own death" and he recommends to "go to the end".

Soler (Chapter "surgery"), warns against the potential psychological dangers for trainees, between the trivialization of the death of the patient and the potential trauma of the residents. He recommends to articulate the achievement of the ultimate situation with special educational goals.

At the meetings of our working group, the question was long argued of whether or not a scenario should be allowed to evolve up to such a limit, with opinions diverging at the beginning. After long debates, our group agreed to the following conclusions: the achievement of the ultimate situation allows students to confront simulation in a stressful situation for which they will learn, without real risk, to manage their stress and implement strategies for recovery of the degraded situation before the unacceptable; it is better to learn it in simulation than in operating situation or real intervention. This choice also offers trainees the opportunity to acquire knowledge on their own functioning situation "limits", i.e. to acquire meta-knowledge, as pointed out by Amalberti (1996) about the importance in the setting of "cognitive compromise" between the operator resources, the current problem treatment and the tempo of situation evolution.

The achievement of ultimate situations must also be worked in recalling elements of the didactic contract (the pedagogical objectives, the contribution for the trainees) and in preparing the accompaniment of trainees, as well as trainers. The management of the situation and of the post-situation must be previously thought. One must particularly wonder about how to lead trainees to benefit from dealing the situation on simulator: it is not recommended to let trainees leave the training session after such a failure without having them to understand any benefit that may be obtained from the confrontation to such a difficulty. In addition, replaying the situation with a successful end before finishing the training session in highly recommended.

1.3 The Simulation Situation

The simulation situation can be described by three times (Pastré 1997) that will be used to structure our comments: the briefing, going onto the simulator, the debriefing.

1.3.1 The Briefing

Samurçay and Rogalski (1998) indicate that the briefing is the opportunity for the trainer to transmit their operational knowledge useful to the realization of the simulation. Trainers can for example ask trainees to prepare the topics to review the knowledge involved in the simulation and the key points (Chapter "aircraft").

The briefing is also a special moment to negotiate the didactic contract (Samurçay and Rogalski 1998), i.e. the contract between the trainees and the trainers, for which a part remains implicit, and in which the trainers' expectations and the objectives of the simulation are included. We gave an example above in referring to the transposition of ultimate situations.

To make more explicit the didactic contract, some trainers ask the trainees to prepare, about the chosen theme, what they expect practically from the training session (Chapter "rolling mills"). Making the trainees thinking and making them saying what they expect of the training session is indeed not unnecessary. Sometimes, some trainees arrive in the session without any expectation (at least without explicit expectation), others with expectations shifted from the proposed pedagogical objectives. This prior discussion permits on one hand a more constructive end-of-session, and on the other hand contributes to closer compromise that each trainee is called on to do with regards to different types of constraints to which s/he must confront (we shall develop it later).

In the perspective of didactic contract elaboration, Fauquet and Frémaux (2007, 2008) have listed for anesthesia training a set of points to be discussed in an attempt to establish a relationship of trust with the trainees and avoid that they feel trapped or put in difficulty voluntarily by trainers:

- clarifying techniques for the use of the simulator (for example osculation points), with a proposal for the trainees (and people attending the course to some extent) to come closer to the mannequin simulator to see and touch while time explanations are given;
- defining the pedagogical inputs of the simulator (training to practice the gesture, to perform medical analysis);
- remembering that the simulator permits the discovery of rare situations;
- clarifying that the simulator will allow them to work interaction and communication in the team, leadership, management of the environment;
- remembering that training is not an exam, thus without any evaluation of the trainee (specify ethics);

- describing the security dimension (no risk for the patient);
- discussing what the objectives of session are and what is not the session (in particular, the trainers are not there to trap them: recall that situations are constructed from real experiences in operating or intervention).

1.3.2 Going onto the Simulator

When the simulator session begins, the trainer must carry out a multitude of tasks, as had also noted it Antolin-Glenn (2005) for the Fire Department officers training. For the reactor or aircraft pilot training, the trainer operates the simulator, and also simulates roles of third parties not present in a situation of simulation (maintenance staff, air traffic controllers), ensures the conduct of the scenario, observes the trainees individually and collectively. Also permanently, s/he adapts the evolution of the situation to the trainees' actions. For this aim, s/he has knowledge of the trainees' profession since s/he must be able "to make positively changing the evolution of the scenario if the management was finally correct, even if different from that described in algorithms (Chapter "anesthesia").

In this frame of involvement in the situation, the place of the trainer is both enveloping and inserted in the simulation situation: enveloping because s/he conducts the situation, by controlling or adjusting the parameters of the simulator or the answers s/he gives to the reactor pilots based on the role s/he must play (the maintenance technician or the field worker for example), and inserted precisely because these contributions take place in the history of the situation simulated by temporal intervals. The trainer's position is enveloping also by the distant observer position which will be essential in the management of the debriefing.

Trainer undertakes a permanent adjustment of the situation of simulation to the trainees: it is to maintain them in their zone of proximal developmental (Vygotsky 1934/1986). The simulated situation is in fact the trainer's provided plan, plan to be adapted to the progress of the situation of simulation. Any trainer is a designer: s/he designed and prescribed tasks for individuals in training (Faita 2003), and in situation s/he must adapt this "preconceived" to the activity of the trained subjects as Altet (1994) perfectly showed it for school education. In simulations by which we are interested here, the aim is to keep trainees in problem-solving situation "which are professional situations where an operator is obliged to transform her/his repertoire of skills to adapt it to a new professional condition" (Pastré 2005).

Insert 1 The zone of proximal development
Schematically, it is considered that the zone of proximal development can be defined as the difference between what the subject knows to do with the help of others and what s/he knows to do independently. In this line of thought, Bruner (2002) has defined the guardianship [one could say the way other who

knows, the trainer, is helping people who are not able to act autonomously] and has characterized various modes of intervention of the trainer.

At the same time, the trainer can intervene to help resolve the problem. To prove this, we can rely on the contributions of Bruner in the line of Vygotsky who defined guardianship, as "the means by which [...] a specialist is helping someone who is less [...] a specialist that him/herself" (2002, p. 261).

By analyzing vessel conducting training on simulator, Percier and Wageman (2004) identified three forms of regulation in the activity of guardianship of trainers:

- direct regulations: they accompany step by step the realization of the trainees' actions specifying the action patterns, the controls to make, etc. This form of guidance is little autonomy for the trainee and avoids the production of errors;
- proactive regulations: in setting out the task, the trainer provides early certain ways of explanations and information (such as the operating modes, risks, etc.). This guidance mode allows the updating of the trainee's representation. When errors are committed, as soon as the trainer detects it, generally, the intervention takes the form of retro-active regulations.
- retro-active regulations: they outline the task and observe its completion by the trainee. When performance deviates from what is expected, the trainer may advise a remedial action, attract attention and propose a solution, explain, or also assess.

These forms of regulation are to be in connection with the didactic and tutorial choices. Thus for example, in a study of interventions by the trainers to driving, Vidal-Gomel et al. (2008) noted, in the same way as Percier and Wageman (2004), that the direct regulation seems rather intervening at the beginning of training. Didactic and tutorial choices are also made, implicitly or explicitly, in the interventions in case of a trainee's error. Percier and Wageman (2004) have gathered some of the main questions:

- Is it necessary to intervene after the completion of an error?
- When to intervene? Immediately after? If yes, it does not allow the trainee to understand the immediate consequences of the error in the action. Later? Then taking the risk of an accumulation of errors, which makes them more difficult to analyze afterwards, and is also a risk of a loss of control of the system?
- How to intervene? The simulation can be stopped, interventions may be more guiding (orders for completion of action are given).

The analysis of guardianship may also be a good indicator of the implicit or explicit objectives of trainers. For example, Vidal-Gomel et al. (2008) became interested in the interventions of five trainers in a trip-test performed by older drivers. They have highlighted two different types of guidance (in terms of form, content and timing of the intervention) and two invariant guiding organizations: patterns, indicative of the more or less explicit objectives of trainers who participate in these formations. Some seem to just have the objective of making the point with the trainees while others also seek to transform the driving behavior. These ones give a more ambitious objective than the one announced for these formations.

The whole work activity of the trainer is not so simple at this stage of the simulation, because at this stage, s/he must also gather the matter s/he will use in debriefing. However, as noticed Fauquet-Alekhine and Maridonneau (Chapter "reactors"), it is often when the activity is the busiest for the team of trainees that the trainer is the most requested on operational settings and interactions with the simulator and the trainees, while in parallel it is necessary to collect evidences of the activity to prepare the debriefing.

To gather these trails, taking notes, and, when possible, sharing tasks in real time between trainers, become a factor of success of the conduct of the debriefing. Labrucherie advocates that the trainers, in the preparation of the session, must identify what are phases of scenario which will induce to such and such behavior of the trainees, to anticipate increased observation phases. The observation of the trainer must be "specialized", focused, first to remain effective in relation to the final expectation (for this aim, see for example observation grids detailed in the appendix to the chapter regarding aircraft piloting, including a grid centered on stress, which defines a few "observables").

Regarding these "focused" observations, Labrucherie (Chapter 'aircraft') notes that it is difficult to train trainers to focus observation on human and organizational dimensions, or, in other words, to observations of the collective activity. This remark should probably be close to the work of Rogalski et al. (2002) which showed the lack of interventions by the trainers to the collective dimensions of activity both in aircraft pilots' training and in training for collective decision-making of firemen officers. It appears thus that analyzing the collective activity in situation and guiding it could constitute a real difficulty.

The observation should be focused, but not only. The trainer must also be able to capture, outside these predefined items, elements that must be not discussed otherwise s/he might let the trainees go away with ways to do or not to do. One of the most common examples encountered by trainees is the focus on a particular problem which led to hide other elements of the situation which would have required special treatment or provided information to better understand the situation (see Chapter "reactors" for example). Catching this kind of events is all the more interesting as it deals specifically with how to perform the profession, individually and collectively, and as it questions the cognitive compromise drawn up by the operator (within the meaning of Amalberti 1996). These are indeed the objectives of training through simulations, such as those presented in the present book.

Another example of what they can offer to the team is observing and sharing the "tips", the professional "tricks" that everyone knows how to implement to facilitating or improving the work. Chapter "reactors" gives two examples: that of the sheet of paper that the pilot places on the control panel to make sure not manipulate those buttons who are hidden, and that of the pilot leaving open the hood of a recorder to be sure not to omit an important parameter tracking from time to time. These "tricks" are the deviation of the habitual use of an object to another use, and are named "catachresis" (Faverge 1970; Rabardel 1995; Clot and Gori 2003). It can also be analyzed as the anticipated management of one's own errors, which is named by Amalberti (1996) the internal risk.

To assist the trainer in these observations, the video can be a relevant tool (Fauquet-Alekhine and Maridonneau 2012). In conditions at least:

- that the viewing of video sequences in debriefing is not at the expense of the development discussion (there must be only illustrative support);
- that the students feel no loss of dignity (see a critical example in Chapter "reactors" and Fauquet-Alekhine and Maridonneau 2012).

Ideally, the video system must allow to show the activity of work, and not just the trainees, and must be equipped with a device for quick search of the relevant sequences (because identification and screening of sequences to analyze in the debriefing should not be done at the expense of the discussion).

Other types of records are possible: those of the evolution of the physical, technological, or physiological parameters as appropriate, that trainees can observe in real operating situation or real intervention, but that the simulator via its calculator allows to materialize on graphs. Yet this is one of the benefits of the simulator: give access to these usually invisible parameters for help to a better understanding of the process and the effects of one's actions. However, as for the video, the use of these records must be made sparingly, otherwise it will bog down debates in purely technical details (Bonavia (Chapter "rolling mills")) as well as Fauquet-Alekhine and Maridonneau (Chapter "reactors") comment this point).

A transition between going onto the simulator and the debriefing should be planned; a break is necessary, and if possible, long enough with regards to the time spent on the simulator. For example, nuclear reactor and airline pilots spend several hours on simulator, and the break is 30 min.

In addition to recovery, this break has several virtues for trainees:

- It allows them to do something else, and therefore take a little distance from the operator in action.
- It is an opportunity to possibly discuss what they lived; thus, the distance is made through the spontaneous "putting into words" they do,
- It takes out of the future discussions considerations too much oriented on technical aspects of the process: trainees usually start by interacting between them on this type of question; what is stated during the break on these themes will be in debriefing of most synthetic way, leaving more place for discussions about professional practices.

For the trainers, the break allows to update notes and aspects that appear to be unavoidable in the discussion to follow, to elaborate possibly a strategy to lead the debriefing and, if there are several trainers, to confront the views of each on what they will do.

1.3.3 The Debriefing

The importance of the debriefing is such that none of the co-authors of the present book describes training on simulator without it. However, the matter discussed in debriefing and the qualities of this matter are depending on what happens on simulator.

We will mention here a few essential aspects of what the co-authors of the present book have pointed out about the debriefing: leading the debriefing, managing the stress, the "simulator effect", and the place and functions of the failure. In doing so, discussions on these four points will not be balanced: we will discuss the guidance of the debriefing in more details whereas the other three points will only be mentioned.

- Leading the debriefing.
  After the run on the simulator and the break, trainees go in the room for a debriefing.
  One of the fundamental questions concerns which length of time must be granted to this period of work. The experience of the authors of this chapter led to recommend at least equal to the time spent on simulator because, as we shall see, the collective discussion of the professional development is neither spontaneous nor obvious for trainees. In addition, other elements must be put into discussion to create a relationship of trust (Labrucherie, Chapter "aircraft") or even to treat potential interpersonal conflicts, as told by Fauquet-Alekhine and Maridonneau (Chapter "reactors") or Labrucherie.
  The relationship of trust is not established at once because, even if trainees and trainers know each other previously, it is not uncommon that trainees have still in mind that the trainer is there to "trap" them. So the bias must be "demystified". Reminding the pedagogical objectives, and beginning with a round table to know and then identify the expectations of everyone in the training, can help. At the debriefing, it seems worth reminding the didactic contract.
- Regarding the possible interpersonal conflicts, Fauquet-Alekhine and Maridonneau (Chapter "reactors") related an example in which 20 min from the time of debriefing were devoted to the management of conflicts. They argue that if this time is not taken, trainees are not fully receptive to the work of debriefing about the run on simulator. Labrucherie exposes a case where the debriefing lasted 30 min, after which one of the trainees discussed with him for one additional hour to understand what had happened and to continue the work of transformation of his style (within the meaning of Clot 1999; see Insert 2).

As we shall see through the following chapters, the debriefing is a development discussion of the profession based on what was lived during the simulation situation. Pastré (1999, 2000) speaks about "retrodiction" to designate it. In each of the debriefings, people in their retrodiction, go through various evolutionary phases of the content of their comments (Béguin and Pastré 2002; Pastré 2005; Fauquet 2006; Fauquet-Alekhine and Labrucherie 2012). The first transformation is to transform the experience of the simulated situation in a story. Gradually, this story is structured by ordering the elements in a unit of meaning, from the story to the intrigue. Finally, with explanations to better understand the intrigue, the collective manages to the ultimate phase sought and consisting of the development of knowledge.

In a Piagetist perspective seeking to promote awareness (Piaget 1974), "it is better to make trainees speak rather than let trainers lecture" (Bonavia, Chapter "rolling mills"), with caution: among the trainees, the trainer must always question the hierarchical subordinate first "to avoid the superior's influence" (Labrucherie, Chapter "aircraft").

Based on the conceptual approach proposed by one of the perspectives of French psychology, the clinical analysis of the work activity (Clot 1999; Clot et al. 2000; Fauquet 2006; Fauquet-Alekhine and Labrucherie 2012), we can highlight the importance of implementing the discussion of the workers' action by themselves—see Fauquet (2006) for the use of this approach in other contexts. They are asked to explain what they are doing beyond of what is a priori agreed, to re-formulate—as in a more classical self-confrontation analysis (Theureau 1992; Mollo and Falzon 2004)—but also to understand the way in which each one is approaching the situation beyond what is agreed a priori and might remain implicit.

The debriefing, considered in this framework, aims to extend the implementation discussion beyond the implicit within the story, suggesting that the development activity is governed by conflicts between concurrent activities that may be incurred for a same task to achieve but with different costs (Clot 1999), which is a specific viewpoint of the crossed self-confrontation practiced in the clinical activity. The activity carried out must be put into discussion, but also suspended activities, thwarted or affected activities, and even including of counter-activities. In relying on the collective development of the story, then of the intrigue, the analysis highlights for the workers a lived and shared history of what has built the situation. During this phase of collective discussion, the "cross-confrontation" is implemented for a necessary comparison of the "personal styles" through the "professional genre" and make them to evolve. The context of the simulation situation is therefore reconsidered to allow its re-construction or just its interpretation.

Insert 2 Genre, style, and cross-confrontation (Clot et al. 2000; Fauquet 2006; Fauquet-Alekhine and Labrucherie 2012).

Genre: "Professional genre" is defined as a social dimension where values are shared that regulate tacitly the work activity with regards to formal prescription, informal obligations arising from the collective itself, history and way of life of the collective (including the attitudes and the ways to communicate). The shared rules of the professional genre generate constraints and generally are also a resource for actors insofar as they are not fixed, but can be re-examined and be transformed. To be effective in all of the situations encountered by subjects, professional genre cannot be frozen; it must be re-visited, re-examined permanently to be adapted to the context of work which also changes.

Style: To be effective in all of the situations encountered by subjects, professional genre cannot be frozen; it must be re-visited, re-examined permanently to be adapted to the context of work which also changes. This process is shaped using the professional style of each, and by confronting between them in the professional genre, re-defining it.

Cross-confrontation: to help actors to access to this evolutionary process of the genre and of the style, co-analysis helps to put into discussion style and genre in cross-confrontation, developing the power to act, both personal and collective. During this cross-confrontation, the elements of discussion usually unnoticed in daily life are maintained more obvious to allow their re-work. This is fundamental to the analysis because it leads to act on the development of the subject, the collective, and the work environment.

It is therefore a co-analysis in the collective debriefing which must re-examine the professional genre. The common rules of the professional genre are both constraints and resource for workers insofar as the rules are not fixed, but can be re-examined and processed. The professional genre performs a psychological function for each worker through a transpersonal dimension (Clot et al. 2002; Fauquet-Alekhine and Labrucherie 2012). To be effective in all of the situations encountered by workers, professional genre cannot be frozen; it must be re-visited, re-examined permanently in order to be adapted to the context of work which also changes, and in order to develop knowledge and new skills. This process is shaped by using the professional style of each, and by confronting each other within the professional genre, redefining it through the transpersonal memory (within the meaning of Bannon 2000). Labrucherie (Chapter "aircraft") speaks of "interactive analysis". To help the trainees to gain access to this process, the co-analysis helps to put into discussion style and genre in "cross-confrontation", developing the power to act, both personal and collective. During this cross-confrontation, the elements of discussion usually not noticed in daily life are maintained more obviously to allow their re-working. This is part of the role of the trainer animating the debriefing.

Working the "dialogical residues" (Scheller 2001) is then fundamental because it leads to having an effect on the development of the subject, the collective", and the work context. An example of exploitation of dialogical residue in debriefing is given hereafter: on simulator, a nuclear reactor operator inhibited the klaxon informing the appearance/disappearance of the alarms in control room. His action was motivated by the fact that an electrical defect maintained permanent klaxon, which induced a noise nuisance hard to bear by the piloting team. In addition, this klaxon no longer fulfilled its function; it was therefore useless. Inhibiting the klaxon would allow pilots to work more smoothly but implied that they would implement a monitoring strategy "to view" alarms. All agreed in the action, but involved in multiple tasks, they moved away from the treatment of the alarm for several minutes, which was observed by the trainers. During the debriefing, trainees tended to skip very quickly over the discomfort caused by the klaxon. An operator said: "Yes, and then the klaxon was a bloody mayhem" ("*chienlit*" in French). And the retrodiction continued on other facts. Trainers then interfered to put emphasis on the term "*chienlit*" and asked him to explain how he could describe the situation. The beginning of the questioning allowed trainers to tell the trainees what their motives were, their objectives, what they were planning to do, and what they have done and not done. This analysis led the pilots to realize and understand one of their weaknesses in situation, and suggested what they could do to better manage the situation if it happens again.

Metaphors can be used in the same way. Indeed, one of collective analysis techniques to access a share of the work activity that is not immediately visible, is to take metaphors and "to make workers speak" on their meanings and their motivations. However this is complicated by the involvement of the trainer within the history of the simulated situation, and inside the professional genre. Indeed, seizing metaphors is as much of the speech analysis than of the listening, attentive to what is said, distant and voluntarily "naïve". Concerning the professional genre, the more trainers have experience of the pilot profession, the more these metaphors go unnoticed because they are adopted and because they immediately make sense to them; this requires trainers' awareness to realize, analyze, and eventually transform the metaphors as an object of debate and training.

Silence is another element of dialogue interesting to use, such as highlights Labrucherie (Chapter "aircraft"). Quite rightly, he reminds that "silence is an efficient tool to express a meaning" which "you need to let it last long". Vermersch (1994) provides valuable guidance on the management of the time of silence, to access the implicit potential of the speech by taking into account, in the activity analysis, actors' emotions participating to the explicitation. In addition, this avoids the trainers to put their own words on what the trainees want to express, or to induce their own representation of the situation to them. Vermersch thus evokes the "creation of a new reality."

During the debriefing, implementing the discussion proceeds essentially by questions about the profession. Each participant questions the profession, as when expressions of each actor are introduced into the discussion, it provokes other questions. In this exchange, the trainer's role is to ensure that knowledge and expertise, style and genre, professional practices, are effectively shared, make controversy, are questioned for access to the sought effects.

One of the traps to avoid for the trainer is to stay on the "why?" without going on the "how?".[2] The temptation is great because in general, there was action or lack of action, the interlocutors are likely to say "why do you (shouldn't you) do it?". In taking the why, answers and exchange take the risk of to ponder, or even to focus on what generates the why, i.e. prescription or technical matters. However, if the

---

[2]To go further about interviews and the different kinds of interventions, see Blanchet (1991) and especially Theureau (1992), Clot et al. (2000), Mollo and Falzon (2004) regarding self-confrontation.

debate on the prescription or the technique may be necessary, knowledge and know-how may not be transformed. It is important for the debate to be about the "how". This is what is stated in the Chapter "reactors", proposing to engage by this questioning in analysis of practices and not that of the industrial process technology. This is also discussed in the Chapter "rolling mills".

This approach is particularly effective by questioning the how; it permits to highlight, by and for the trainees, implicit practices or what has been agreed in the activity, or even what the workers are not aware of, either in action or in the absence of action. Indeed, the "why" puts into discussion only what is visible, therefore either what trainees become aware of, or what they do. Yet the work activity is not only that: all the dimensions of the activity are not perceived or "perceivable", and activity also includes what is done and what is not, and what could not be done, and has been prevented. Fauquet-Alekhine and Maridonneau (Chapter "reactors"; see also Fauquet-Alekhine 2012a) illustrate about a mode of communication between pilots that, when implemented, gives proof; but the work activity analysis shows that this only works out for a small number of actors of this "shared repository", which becomes "unreliable" as soon as a third party integrates the team. The problem is identified in this example from the analysis of activity; it may also be based on experience feedback.

The constitution of shared repositories is a guarantee of the reliability of the operation of a collective work (Terssac and Chabaud 1990). When it becomes a too tacit dimension of skills, the collective excludes at once new workers, which can contribute to generate errors, failures, or even accidents. It is therefore important that training contributes to the awareness of this type of implicit.

Labrucherie (in the appendix to Chapter "aircraft") presents a grid of observation focused on situation awareness (in terms of Endsley 1988), which are important for risk management in dynamic situations. On the other hand, Geeraerts and Trabold (Chapter "anaesthetist") report how some reactions of the trainees are sometimes completely ignored by themselves.

In addition to collective speech, the trainer has at disposal some extracts of video recordings on simulator to show some actions carried out by the team for which the trainees are not conscious. For example, it is not uncommon that the team implements reliable practices which seem so natural that they are unable to carry them spontaneously in the debate; yet how disseminate effective practices if the actors themselves are not aware of them? This clearly contributes to anchor these practices in learning.

Thus, the trainer must ensure that the debriefing deals with the real of the activity, not only with the realized, at the risk of not achieving the objective sought. That is why the trainer, distant actor himself from the simulated situation, is essential during the debriefing because he retains a capacity to be surprised at what has become normal for the trainees. His/her astonishment, his/her questioning and analysis will help the trainees to put another look at their actions. This progressive distance contributes to help workers to untie in with what they have done, of what they used to do. The trainer leads pilots to explain what they have done or not done depending on the context, and on the basis of this collective speech, s/he manages to change professional practice in order to stabilize them or make them more reliable by adapting them. It is to guide reflection on the activity and to promote awareness to transform the individual and collective skills.[3]

The debriefing a special moment to deal with stress

For professions depicted in the present book, stress is an important parameter. According to the trainees' level of stress, their performance can be either improved or deteriorated (Yerkes and Dodson 1908; Mclean 1974; Staal 2004; Müller et al. 2009; Jo et al. 2013; Fauquet-Alekhine et al. 2011, 2012, 2014). Sometimes, it gives rise to surprising reactions whose trainees are even not aware of (see Chapter "anesthesia"). According to Labrucherie (Chapter "aircraft"), trainees must be lead to speak about the stress, while taking care to avoid that they lose self-confidence in them or their teammates. He explains that it is important especially to allow the trainees to identify the stress, to talk about it, and build with the trainer the conditions which will allow to lower stress levels towards acceptable conditions.[4] The debriefing is the designated space for this, and the rationalization of the living situation which is allowed here helps the trainees to better manage it the next time and perhaps helps to develop a meta-knowledge, of which we previously noted the importance.

The "simulator effect"

We cannot finish this paragraph on the debriefing without speaking of the "simulator effect" (in French: "*l'effet simu*") sometimes relied on by the trainees. It was only mentioned in Chapter "reactors" but we have experience about it in training simulator with anaesthetists and airline pilots, as well as with the nuclear reactor pilots.

We will leave here aside contextualized aspects already developed elsewhere (Chapter "reactors"; see also Fauquet-Alekhine 2012b, p. 49) to think about a more general meaning of the "simulator effect". Given the type of training to which the present book is interested in (the development and certification of professional practices already acquired by the workers), it seems that the issue could be approached with the skills perspective in these terms: is a skillful professional in real operating situation necessarily skillful on a simulator? Indeed, we explained how a simulation may approach more or less close a reference situation: figurative and operative simulator dimensions can be more or less developed, the variability of the situation may be more or less important, and the scenario can be simplified or not according to the pedagogical objectives. In fact, the simulation situation is not a situation for which the trainee has all the professional skills required (with regard to

---

[3]See also Falzon and Teiger (1995) interested in the ergonomist' role to guide reflexive activities and to favor awareness.

[4]Nevertheless it remains difficult to elaborate a standpoint regarding "acceptable conditions" for the subjects and for their health. In this chapter, we shall only consider that, in risky situations for oneself, the others or the environment which we deal with in the present work, it is necessary for the subjects to monitor the situation enough to stabilize it and avoid the accident. Therefore "acceptable conditions" means that the subjects are able to act efficiently in the situation.

the training situation, otherwise, s/he could not learn anything from the simulated situation). Just as it would be unacceptable to say that a worker recently trained on simulator will be immediately operational and skillful in his/her daily work activity, it does appear to fail to expect from a professional spending most of her/his working time in real operating situation to be skillful on a simulator, especially if the simulation situation appears far away from his/her daily activity. A discussion with physician-trainers of anaesthetists illustrates this point: they explain that in general, if the residents have no problem to engage themselves in simulation training, it is not the case of anesthetists in position for many years. They seem to perceive the situation of simulation as an opportunity to possibly make visible some of their problems while they are role-models for their trade. For them, the simulator is therefore not a tool of progress or improvement, but a conducive object to discredit.

Therefore, it seems to us that the professional worker is at once confronted to difficulties on simulator. This proposal is also perfectly illustrated by experience reported in Chapter "surgery": Soler explains how computer engineers have obtained better results than surgeons during first manipulations on simulator, stressing that a time for "perfectly mastering the system" is needed, "the lack of realism of the scene being the main drawback for surgeons who could not locate their natural landmarks".[5] Also, when a trainee explains that s/he would not do so in non-simulated situation, the assertion cannot be dismissed but must be analyzed.

To reduce this type of "simulator effect", it seems to us that a phase of (re) takeover the simulator by trainees is fundamental. It is divided at least in two steps: the first consists of the introductory phase of the simulation situation with trainees (in the section devoted to the briefing we mentioned it related to the concept of didactic contract) and the second includes the time required for learning of the system (see Soler, Chapter "surgery").

Let us go back to the time spent on the development of didactic contract, which appears all the more important in that it allows to create a relationship of trust with the experienced professionals who may fear to be "trapped" and to lose face towards themselves, trainers or their colleagues. Recent studies with physicians-anesthetists have demonstrated importance and consequences induced by the form it takes, or that the trainers give to it (Fauquet and Frémaux 2007, 2008). The introduction "determines the success of the scenario (in that it may be done whole, reaching the established pedagogical objectives), and a proper understanding of the simulator. For example, during an introduction, it was omitted to clarify that if intubation is difficult, one should not force. That would degrade the material, and in a real operating situation, this could traumatize the patient. 'If this is difficult, it is perhaps that it is expected in the scenario.' When this prerequisite has been omitted in the introduction, the pair of trainees forced intubation and it was successful. The consequence: a simulation break by trainers asking them to consider that intubation was impossible. In practical terms, trainers had to go back

---

[5]This example also reminds to which extent the characteristics of the artifact "simulator" as well as their use by the subjects must be carefully analyzed.

in chronology of the scenario: trainees undo what they succeeded to do because the expected sequence of the scenario did not allow this. The result of such flashbacks is a loss of credibility of the simulated situation, since it forces to undo what has been done" (Fauquet and Frémaux 2007, 2008).

- The failure as a lever of development.
- If we accept that a professional is immediately in trouble during the simulation, then the "failure management" during the debriefing must be considered. It seems that the situation of failure experienced by the trainees cannot be what about they will leave the simulation session. From a pedagogical perspective, the simulator cannot be a demonstration tool of the trainees' "non-knowledge", but must be what allows them to progress. Thus, a simulation session which would be completed on a failure is an unfinished session. Trainers are required to give the trainees the means to transform their knowledge, know-how and practices, individual or collective.

Yet between the total success and failure are intermediate levels which are necessary to evaluate. Geeraerts and Trabold (Chapter "anesthesia") note that assessment can be problematic: who evaluates, what is the legitimacy of the assessor, particularly when trainees and trainers are peers? Should the assessment lead to validation of skills? One might also ask: what do we evaluate? a performance? a progression? For aircraft pilots, the question is a bit different since the readiness assessment is established for a long time, and imposed by the regulator. Yet what are the consequences for the training process? For nuclear reactor pilots, there are two levels of appreciation: an annual assessment that allows the management to appreciate the piloting certification, and a systematic evaluation at the end of training session using a progress help form (in French: *Fiche d'Aide à la Progression*); proposed by the trainer to the trainee and archived by the two parties, it registers axes of progress and helps to guide the content of upcoming training sessions. Thus, the certification and the assessment of what was acquired in training are well differentiated, which is not always the case and leads to new questions. For example, must simulations be developed to assess the operators' ability be also considered as situations of training? Under which conditions these contribute to the training?

In conclusion, it seems to us necessary to have an assessment system of the trainees' progress with the precautions mentioned by Bonavia (Chapter "rolling mills"). Being qualitative or quantitative, such a system is a landmark for all: the trainee knows where s/he is, and the trainer knows which points to improve. It is also a way of knowing whether the training is relevant: a training that would bring no progress serves no purpose, and the only way to access this knowledge is the implementation of an evaluation system. To go further, we may notice that aircraft pilots' trainers have a tool for assessment of their training practices which could likely be an interesting mean to improve the training on simulator. Indeed, why be limited to the idea that only the trainee would have to making progress?

Thus the simulator training draws attention to the problem of the assessment, which is discussed in the field of education (Figari and Mottier-Lopez 2006, for example) but which is still too often absent in the research on the professional training based on analysis of real operating situations.

2 Understanding the Activity in Terms of Antagonistic Goals Which the Trainees Must Face

So far we have been discussing activities or tasks of trainers in reference to frameworks of analysis of the simulation and the different phases of the simulation. In so doing we have pointed what would be good to be done with regards to pedagogical objectives. But are the trainers always able to do so? What are the choices they are forced to do?

The trainer must cope with their company constraints, and their career will likely play an important role in the arbitrations they will have to do. For example, if the pedagogical goals require 2 trainers for 5 trainees and 2 h of debriefing for 2 h of simulator, on the opposite, the financial logic of business can lead to optimize the ratio of trainees/trainers and time spent in training. The trainer who will have to ensure alone the training session will need to find a compromise: a too short debriefing for too ambitious pedagogical objectives will lead to make choices on what will be "quickly said" and what will be seen in depth.

In drawing on issues of ergonomics and ergonomic psychology, the work of the trainer can be analyzed as requiring to develop compromise between contradictory, conflicting goals. This view leads to analyze the work in terms of regulatory activities. "From an ergonomic standpoint, [regulation concept is rather defined with regard to] changes in the behavior of individual and collective of operators to meet the requirements of the situations. Regulation designates the process being implemented by the operators to elaborate compromises between conflicting constraints" (González and Weill-Fassina 2005; see also Fauquet-Alekhine 2012c). Following Leplat's work (2008) based on the issues of the theories of activity, it is considered that the same action can have goals and multiple motives, leading to differentiate between "regulations focused on tasks" and "regulations focused on the subjects" (p. 28). The author gives examples: compromises to be achieved between productivity and security concern regulations focused on tasks, while compromises to be achieved between maintaining health, maintaining good relations with colleagues, etc. are related to regulations focused on the motives. In service activities, Caroly and Weill-Fassina (2007) or González and Weill-Fassina (2005) have clearly shown the need to take into account the two types of regulations. They are part in a model that takes into account 4 opposing poles:

- the pole "system" that includes constraints and resources for the operator, including tasks such as they are defined, or developed tools which are available, etc;

- the pole "other" which corresponds to the hierarchy, to colleagues, the collective at work;
- the pole "self" that is the motives that the operator is seeking to achieve in his/her activity (preserving his/her health, values, personal ethics, etc);
- the person for which the service is intended: the client, the user who must be satisfied.

With regards to this analysis framework, our comments will be more modest as it will be little question of the collective. We consider that the regulation of trainers registers between several poles in tension:

- The pedagogical objectives, which are part of the constraints of the company but are specific to the activity of training.
- The constraints and resources of the system of work, including the company performance constraints.
  If among these constraints, the fact that having high-performing (and therefore well trained) teams is included, there is a long list of other factors which are opposed, such as the aforementioned logic of cost restriction, but also the availability of trainees with regards to their planning of work, the workload of the trainers which sometimes does not allow them for optimum preparation of the training sessions. In other words, under the framework proposed by Caroly and Weill-Fassina (2007) and González and Weill-Fassina (2005), we emphasize here two following objectives opposed to the pole "system".
- The experience, the skills, but also the development of the trainers: as we shall see through the following contributions, the trainers will focus on the areas they know well (according to their life, their professional experience). Trainers who are familiar with the installation may spend much time on this topic even if it is not the main pedagogical objective (see the testimonial in Chapters "reactor" and "aircraft"); trainers who are about to join the operation department may refuse the controversy with trainees recognizing among them a future hierarchical manager (which may be the case of a reactor pilot).
- The skills and duties of trainees: as any educator, trainers must adapt their teaching. They should insert their interventions within the zone of proximal development (Vygotsky 1934/1986) whether contributing in knowledge, implementing professional practice discussion or analyzing the work activity in debriefing.

Figure 2 presents a compromise which would give as much weight to each pole: compromise is schematically located in the centre of the pattern, and gives equal weight to each pole. If one of the poles would be more important from the trainer's point of view, the compromise would move to these poles.

For the demonstration, let us take the example of a trainer forced to intervene alone while two trainers are needed, and in a situation of professional transition: s/he will be soon under the order of one of the trainees (it may be the case of a

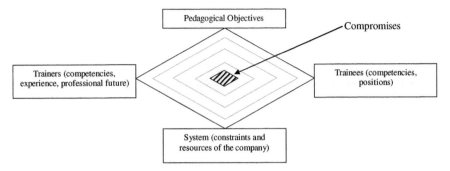

Fig. 2 Four poles "in tension" between which the trainer must establish a compromise

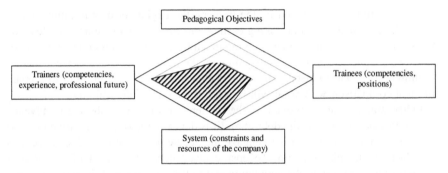

Fig. 3 Schematic diagram of a compromise leading to favor the poles trainers and system

reactor pilot). S/He shifts the compromise to the trainer's development and the constraints of performance (Fig. 3).

Note that the trainer is not the only one in having to develop compromises between goals and antagonistic motives. This is of course also the case of trainees. The schematic diagram that we proposed (Fig. 2) permits also to account of that in considering the more important weight that can be granted to some possible poles at the expense of others. Suppose that the trainees have a high expectation on a particular aspect of their operating activity for which they wish to improve their individual and collective practices on simulator. They favor the pole Trainees and the pole System, possibly at the expense of the original pedagogical objectives of the simulation training (Fig. 4).

It is then up to trainers to develop a more acceptable compromise with regards to their own goals and motives and those of the training.

The elaborative work of these compromises may lead to the relationship of trust about which speaks Labrucherie in Chapter "aircraft" in order to "take the most out of what will follow" in the simulation. In an opposite way, these differences and shifts may constitute an obstacle to the process of training.

This work must be carried out at the first contact between trainers and trainees (Labrucherie). They reveal what we called earlier the "didactic contract." Thus,

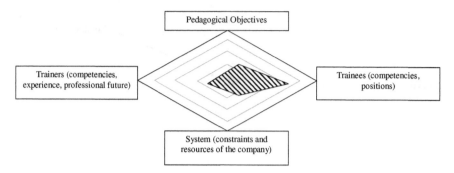

Fig. 4 Schematic diagram of a compromise leading to favor the poles trainees and system

among the strategies deployed by the trainers, the beginning of a training session can be fundamental. It can be asking each one to expose "who we are", to describe one's career[6] (by avoiding too devote too much time to the professional expertise of each: see Chapter "reactors"), to explain why the trainees are together (see Chapter "rolling mills"), and to discuss everyone's expectations a priori in order to prepare the final comparison whit what will be said at the end of the training.

Other strategies are deployed by the trainers: all of them elaborate in advance some particular ways start the debriefing (it is the example of pilot training classes) with a common point which is to engage the debate on the simulated situation and work activity in this situation. To ensure that "technical" aspect of the process remains only a support to the simulation and to the discussion of the work activity, some of them identify priority axes of observation during the run on the simulator (see the four axes in Chapter "reactor").

Thus, when the goals and the motives for the interacting actors in the simulation are shifted, strategies and know-how developed by the trainer within his/her interventions and during the sharing of experience with peers take importance in all phases of the simulation. In addition, as Bonavia noticed (Chapter "rolling mills"), it must be kept in mind that the interactions between trainers and trainees always lead to give the trainers "relevant information for their own job".

3   Training and Development of Trainers

In the present chapter we wanted to make a few theoretical lighting to provide a few additional understanding keys on the trainers' practice. In doing so, it remains to emphasize that we know little about the teaching activity, about the required skills

---

[6]Depending on the topic, explanations may be useless sometimes; for example, trainers for surgeons and anesthetists are specialized physicians, often experienced professionals or professors, facing a population of residents.

and finally about what would be necessary to train the trainers. We will conclude in trying to bring some elements for reflection on this subject and pointing out things to think about for future work.

Trainers can be from operational departments (this is the case of training sessions for "nuclear reactors" or "rolling mills"). This job may also be split between training and operating (this is the case of the aircraft pilots, surgeons or anesthetists). Sometimes, their knowledge is directly derived from their experience of the practice of the profession they have to teach. For others, it involves specific scientific and technical training for several months (or years for physicians or pilots).

In the early days of the training curricula, the use of simulators gives rise to an initial learning on professional technical gestures. This is why trainers are most of the time from the "operational" teams. To give an insight of the following chapters, we may notice for example that:

- rare are the trainers for the aircraft pilots who are not also captain, and rare are the trainers for nuclear reactors that are not coming from the operational departments;
- surgeons' trainers are all surgeons and anesthetists' trainers are all anesthetists;
- trainers for the operators of rolling mills are in pairs, with an operator "referent" of the trade.

Their operational experience gives them some legitimacy to the trainees, or even in some cases an actual legitimacy... But what about the educational component of their work of trainer? Indeed, assuming a good pilot is a good trainer without processing skills is often doomed to failure (see Chapters "reactors" and "aircraft"). Some may claim that the training of trainers has meaning at the condition that the trainer is essential, but Soler (Chapter "surgery") answers to this question by showing that training on standalone simulator is not satisfactory.

In other words, it seems that the issues of training of trainers taking into account the content of the training to be provided are a theme not enough examined when it concerns the training of adults, moreover as it is part of a tradition that recognizes the "work analysis for the training", as had pointed in Montmollin in 1974 (see de Montmollin 1997). Thus, for example we can see that there is relatively little work about professional trainers while lots of researches are developed about the teaching practices in the school field.

The analysis of the teaching "practices" covers a broad set of works that have been developed over the 90s (Marcel et al. 2002). Among this set, a line of research analyzes teaching activity as a job based on ergonomics and ergonomic psychology issues (Amigues 2002, 2003, Bressoux et al. 2009), which is consistent with all of the elements that we have previously pointed out in the present chapter. These researches are a basis on which to rely to better understand what is the activity of teaching, in a training objective, by paying attention to what differentiates the teaching in a school or in professional framework in order to understand the consequences of these differences on the work activity. For example:

- the subjects in training are children or teenagers versus professional adults: the learning difficulties of children and teenagers are not those of adults, their development poses certain problems that a professional will not face, etc.
- the purpose of the training is academic knowledge versus professional skills: knowledge of reference is identified in the first case, this is not always true for professional training; the transposition of this knowledge does not similarly operate as we have previously pointed out about simulations; training professionals does not only enable them to acquire knowledge, but also professional gestures or rules, etc.
- the trainers may have to manage a classroom and to balance the relationship class/individual versus a simulation situation, which involves managing the simulation situation so that it does not deviate too much from the planned scenario, from the "played" roles, etc.

These differences should be studied in-depth and studied especially regarding on the strategies of regulation and the compromises necessary to the realization of the training, with the aim at improving our knowledge on what is the activity of training in professional sectors and the skills required of course, but also in order to improve the training of these professionals.

The objective of the present chapter is more modest. Based on what is said by the professionals of simulation training, we have tried to suggest a few theoretical inputs and emphasize a few directions for future research, to contribute to the objective of the present book: to engage readers to think about their own practices regarding the different testimonials and analyses.

References

Altet, M. (1994). *La formation professionnelle des enseignants*. Paris: PUF.

Amalberti, R. (1996). *La conduite des systèmes à risque*. Paris: PUF.

Amigues, R. (2002). L'enseignement comme un travail. In P. Bressoux (Dir.), *Stratégies de l'enseignant et situations d'interaction. Note de synthèse pour Cognitique, programme «Ecole et sciences cognitives»* (pp. 199–214). Paris: CNRS http://archive-edutice.ccsd.cnrs.fr/docs/00/01/17/90/PDF/Bressoux.pdf.

Amigues, R. (2003). Pour une approche ergonomique de l'activité enseignante. *Skhôlé, hors-série, 1*, 5–16.

Antolin-Glenn, P. (2005). *Le développement des compétences à la coopération dans la formation à la prise de décision distribuée hiérarchisée. Le cas de la formation continue à la gestion opérationnelle des officiers supérieurs sapeurs-pompiers*. Thèse de psychologie ergonomique. Saint-Denis: Université Paris 8.

Bannon, L. (2000). Towards artificial memories? *Le Travail Humain, 63*, 277–285.

Béguin, P., & Pastré, P. (2002). Working, learning and design through simulation. In *XIe European Conference on Cognitive Ergonomics: Cognition, Culture and Design, Catalina, Italy* (pp. 5–13).

Blanchet, A. (1991). *Dire et faire dire*. Paris: Dunod.

Bressoux, P., Kramarz, F., & Prost, C. (2009). Teachers' Training, Class Size and Students' Outcomes: Learning from Administrative Forecasting Mistakes*. *The Economic Journal, 119* (536), 540–561.

Bruner, J. S. (2002). *Le développement de l'enfant. Savoir faire. Savoir dire.* PUF (1ère édition 1983): Paris.

Caens-Martin, S. (2005). Concevoir un simulateur pour apprendre à gérer un système vivant à des fins de production: La taille de la vigne. In P. Pastré (Ed.), *Apprendre par la simulation. De l'analyse du travail aux apprentissages professionnels* (pp. 81–106). Toulouse: Octarès.

Caroly, S., & Weill-Fassina, A. (2007). En quoi différentes approches de l'activité collective des relations de services interrogent la pluralité des modèles de l'activité en ergonomie? *@ctivités, 4*(1), 85–98. http://www.activites.org/v4n1/v4n1.pdf.

Clot, Y. (1999). *La fonction psychologique du travail.* Paris, France: PUF.

Clot, Y., & Gori, R. (Eds). (2003). *Catachrèse: Éloge du détournement.* Presse Universitaire de Nancy.

Clot, Y., Faïta, D., Fernandez, G., & Scheller, L. (2000). Entretiens en autoconfrontation croisée: Une méthode en clinique de l'activité. *Pistes, 2*(1). http://pettnt/v2n1/articles/v2n1a3.htm.

Clot, Y., Fernandez, G., & Carles, L. (2002). Crossed selfconfrontation in the clinic of activity. In *Proceedings of the 11th European Conference on Congnitive Ergonomics, Catalina, Italia* (pp. 13–18).

de Montmollin, M. (1997). *Sur le travail. Choix de textes (1967–1997).* Toulouse: Octarès.

Endsley, M. R. (1988). Toward a theory of situation awareness in dynamic systems. *Human Factors, 37*(1), 32–64.

Faïta, D. (2003). Apport des sciences du travail à l'analyse des activités enseignantes. *Skhôlé, hors-série, 1,* 17–23.

Falzon, P., & Teiger, C. (1995). Construire l'activité. *Performances Humaines et Techniques, n^o hors-série* (septembre), 34–40.

Fauquet, Ph. (2006) Confrontation croisée ou analyse collective sur la base de restitutions d'entretiens individuels: Deux approches pour l'analyse évènementielle. *Revue électronique @ctivités, 3*(2).

Fauquet-Alekhine, Ph. (2010). Use of simulator training for the study of operational communication—the case of pilots of French nuclear reactors: Reinforcement of reliability. *Annual Simulation Symposium (ANSS 2010), 43,* 216–221.

Fauquet-Alekhine, Ph. (2011). Human or avatar: Psychological dimensions on full scope, hybrid, and virtual reality simulators. In *Proceedings of the Serious Games and Simulation Workshop, Paris* (pp. 22–36).

Fauquet-Alekhine, Ph. (2012a). Use of simulator training for the study of operational communication—the case of pilots of French nuclear reactors: Reinforcement of reliability. In Ph. Fauquet-Alekhine (Ed.), *Socio-organizational factors for safe nuclear operation* (1st ed., pp. 84–87). Montagret: Larsen Science.

Fauquet-Alekhine, Ph. (2012b). Causes and consequences: Two dimensional spaces to fully describe short term occupational stress. In Ph. Fauquet-Alekhine (Ed.), *Socio-organizational factors for safe nuclear operation* (1st ed., pp. 45–52). Montagret: Larsen Science.

Fauquet-Alekhine, Ph. (2012c). Training activity simulations: the complementarities of clinical approach and regulation approach. In Ph. Fauquet-Alekhine (Ed.), *Socio-organizational factors for safe nuclear operation* (1st ed., pp. 75–78). Montagret: Larsen Science.

Fauquet-Alekhine, Ph., & Boucherand, A. (2015). Optimal protocol for debriefing of simulation training session (submitted to *Simulation and Gaming*).

Fauquet, Ph., & Frémaux, L. (2007). Simulateur d'Anesthésie – Réanimation. Analyse des pratiques des médecins – formateurs. Congrès MAPAR 2007. Rapport d'intervention du Laboratoire de Recherche pour les Sciences de l'Energie. Référence: LA/./rapint-anrea01 ind00.

Fauquet, Ph., & Frémaux, L. (2008). Simulateur d'Anesthésie – Réanimation. Formation universitaire 2008. Rapport d'intervention du Laboratoire de Recherche pour les Sciences de l'Energie. Référence: LA/./rapint-anrea02 ind00.

Fauquet-Alekhine, Ph., & Labrucherie, M. (2012). Simulation training debriefing as a work activity analysis tool: The case of nuclear reactors pilots and civil aircraft pilots. In Ph.

Fauquet-Alekhine (Ed.), *Socio-organizational factors for safe nuclear operation* (1st ed., pp. 79–83). Montagret: Larsen Science.

Fauquet-Alekhine, Ph., & Maridonneau, C. (2012). Using audio-video recording on simulator training sessions: advantages, drawbacks, and dangers. In Ph. Fauquet-Alekhine (Ed.), *Socio-Organizational factors for safe nuclear operation* (1st ed., pp. 94–97). Montagret: Larsen Science.

Fauquet-Alekhine, Ph., Frémaux, L., & Geeraerts, Th. (2011). Cognitive disorder and professional development by training: Comparison of simulator sessions for anaesthetists and for nuclear reactor pilots. In *Proceedings of the the XVe European Conf. on Developmental Psychology, Pianoro (Italia): Medimond Srl.* (pp. 83–87).

Fauquet-Alekhine, Ph., Geeraerts, Th., & Rouillac, L. (2012). Improving simulation training: Anesthetists versus nuclear reactor pilots. In Ph. Fauquet-Alekhine (Ed.), *Socio-organizational factors for safe nuclear operation* (1st ed., pp. 32–44). Montagret: Larsen Science.

Fauquet-Alekhine, Ph., Geeraerts, Th., & Rouillac, L. (2014). Characterization of anesthetists' behavior during simulation training: Performance versus stress achieving medical tasks with or without physical effort. *Psychology and Social Behavior Research, 2*(2), 20–28.

Faverge, J.-M. (1970). L'homme agent d'infiabilité et de fiabilité du processus industriel. *Ergonomics, 13*(3), 301–327.

Figari, G., & Mottier-Lopez, L. (Eds.). (2006). *Recherches sur l'évaluation en éducation. Problématiques, méthodologie et épistémologie.* Paris: L'Harmattan.

González, R., & Weill-Fassina, F. (2005). Modalités de régulation du processus de travail dans les activités de service en crèche. *@ctivités, 2*(2), 2–23. http://www.activites.org/v2n2/gonzales.pdf.

Hoc, J.-M. (2001). Toward ecological validity of research in cognitive ergonomics. *Theoretical Issues in Ergonomics Science, 2*(3), 278–288.

Jo, N. Y., Lee, K. C., & Lee, D. S. (2013). Computer-mediated task performance under stress and non-stress conditions: Emphasis on physiological approaches. *Digital Creativity, 32*, 15–27.

Laborde, C., & Pastré, P. (2005). *Activités et formations professionnelles: Simulations informatiques comme aide à la conceptualisation. Rapport final.* Avril 2005. CNAM-Université Joseph Fourrier-IUFM de Grenoble.

Leplat, J. (2008). *Repères pour l'analyse de l'activité en ergonomie.* Paris: PUF.

Marcel, F., Olry, P., Rothier-Bautzer, & Sonntag, M. (2002). Les pratiques comme objet d'analyse. *Revue Française de Pédagogie, 138*, 7–36.

Mayen, P., & Vidal-Gomel, C. (2005). Conception, formation et développement des règles au travail. In P. Rabardel & P. Pastré (Eds.), *Modèle du sujet pour la conception Dialectiques activités développement* (pp. 109–128). Toulouse: Octarès Editions.

Mclean, A. (1974). Concepts of occupational stress. In A. Mclean (Ed.), *occupational stress* (pp. 3–14). Springfield, Illinois: Thomas.

Mellet d'Huart, D., & Michel, G. (2005). Faciliter les apprentissages avec la réalité virtuelle. In *Apprendre par la simulation – De l'analyse du travail aux apprentissages professionnels (sous la direction de P. Pastré, ouvrage collectif de l'association ECRIN).* Toulouse: Octarès (Coll. Formation).

Mollo, V., & Falzon, P. (2004). Auto- and allo-confrontation as tools for reflective activities. *Applied Ergonomics, 35*(6), 531–540.

Müller, M., Hänsel, M., Fichtner, A., Hardt, F., Weber, S., Kirschbaum, C., et al. (2009). Excellence in performance and stress reduction during two different full scale simulator training courses: A pilot study. *Resuscitation, 80*(8), 919–924.

Nyssen, A.-S. (2005). Simulateurs dans le domaine de l'anesthésie. Étude et réflexion sur les notions de validité et de fidélité. In *Apprendre par la simulation – De l'analyse du travail aux apprentissages professionnels (sous la direction de P. Pastré, ouvrage collectif de l'association ECRIN).* Toulouse: Octarès (Coll. Formation).

Pastré, P. (1997). Didactique professionnelle et développement. *Psychologie Française, 42*(1), 89–100.

Pastré, P. (1999). La conceptualisation dans l'action: Bilan et nouvelles perspectives. *Éducation Permanente, 139*, 13–35.

Pastré, P. (2000). *Conceptualisation et herméneutique: À propos d'une sémantique de l'action Signification, sens, formation* (pp. 45–60). Paris: PUF.

Pastré, P. (2005). Apprendre par résolution de problème: Le rôle de la simulation, in *Apprendre par la simulation – De l'analyse du travail aux apprentissages professionnels (sous la direction de P. Pastré, ouvrage collectif de l'association ECRIN)*. Toulouse: Octarès (Coll. Formation).

Patrick, J. (1992). *Training research and practice*. London: Academic Press.

Percier, M., & Wageman, L. (2004). Étude de l'acquisition d'une compétence en conduite de processus: comparaison entre deux systèmes d'aide. In R. Samurçay, & P. Pastré (Eds.), *Recherches en didactique professionnelle* (pp. 85–108). Toulouse: Octarès.

Piaget, J. (1974). *La prise de conscience*. Paris: PUF.

Rabardel, P. (1995). *Les hommes et les technologies. Approche cognitive des instruments contemporains*. Paris: Armand Colin.

Rogalski, J., Plat, M., & Antolin-Glenn, P. (2002). Training for collective compétence in rare and unpredictable situations. In N. Boreham, R. Samurçay, & M. Fischer (Eds.), *Work process knowledge* (pp. 134–147). London: Routledge.

Samurçay, R. (2005). Concevoir des situations simulées pour la formation professionnelle: Une approche didactique. In P. Pastré (2005). *Apprendre par la simulation. De l'analyse du travail aux apprentissages professionnels* (pp. 221–239). Toulouse: Octarès.

Samurçay, R., & Rogalski, J. (1998). Exploitation didactique des situations de simulation. *Le Travail Humain, 61*(4), 333–339.

Scheller, L. (2001). Les résidus des dialogues professionnels. In *Education permanente, clinique de l'activité et pouvoir d'agir, Paris, France* (Vol. 146, pp. 51–58).

Staal, M. (2004). Stress, cognition, and human performance: A literature review and conceptual framework. NASA report, reference: NASA/TM—2004-212824.

Terssac (de), G., & Chabaud, C. (1990). Référentiel opératif commun et fiabilité. In J. Leplat & G. de Terssac (Eds.), *Les facteurs humains de la fiabilité dans les systèmes complexes* (pp. 111–139). Marseilles: Octarès Entreprise.

Theureau, J. (1992). *Le cours d'action: analyse sémio-logique. Essai d'une anthropologie cognitive située*. Bern: Peter Lang.

Vermersch, P. (1994). *L'entretien d'explicitation*. Issy-les-Moulineaux: ESF Editeur.

Vidal, R., Frerson, Ch., & Jorda, L. (2011). Training incident management teams to the unexpected: The benefits of simulation platforms and serious games. In *Proceedings of the Serious Games and Simulation Workshop, Paris* (pp. 43–48).

Vidal-Gomel, C. (2005). Une approche préalable à l'analyse de l'activité et des compétences. *Éducation Permanente, 165*, 47–68.

Vidal-Gomel, C., Boccara, V., Rogalski, J., & Delhomme, P. (2008). Les activités de guidage des formateurs au cours d'un audit destiné à des conducteurs expérimentés et âgés. *Travail et Apprentissage, 2*, 46–64.

Vygotsky, L. S. (1934/1986). *Thought and language*. Cambridge, MA: The M.I.T. Press.

Yerkes, R. M., & Dodson, J. D. (1908). The relation of strength of stimulus to rapidity of habit-formation. *Journal of Comparative Neurology and Psychology, 18*, 459–482.

Airliners Flying

Marc Labrucherie

1 Introduction

In less than a century, aviation has grown from the pioneer era to the industrial era which is well known nowadays. And it achieved the incredible performance to multiply its security factor by 30 in 30 years and by 100 in 50 years.

Today, civil aviation is experiencing a crash every million and a half of flight hours and when compared to other transportation means, it has become the safest one.

Many factors have contributed to this great performance, but a large contributor to this progress has been and remains the flight simulator.

From the first flight hours in flight school until the end of his career, a pilot trains and practice regularly in simulated situation.

But what is a simulator?

In flying school, pilots first encounter "simulators" to learn instrument flying. Mostly these simulators represent generic cockpits, not a particular type of aircraft.

Then, in a company or in the premises of an aircraft manufacturer, the pilot who qualifies on an airplane type trains on a simulator which perfectly represents the aircraft cockpit.

This is the type of simulation that we will focus on in the present book, in other words, we consider an experienced pilot who qualified on an aircraft type then maintains his qualification.

It is to be noticed that for the same aircraft, e.g. B747-400, there are two levels of simulation:

M. Labrucherie (✉)
NML Consulting, Cassis, France
e-mail: malabrucherie@gmail.com

© Springer International Publishing Switzerland 2016
Ph. Fauquet-Alekhine and N. Pehuet (eds.), *Simulation Training:*
Fundamentals and Applications, DOI 10.1007/978-3-319-19914-6_2

The Fixed Base Simulator, which is the exact reconstruction of the cockpit but has no ability to move, nor visualize the external environment; it is primarily used to study the operation of the circuits and breakdowns; and the Full Flight simulator.

1.1 The Full Flight Simulator

A full scale simulator mounted on jacks, reproducing sensorial illusions of acceleration, deceleration, and turning.

On the windshield, the Outside World (OTW) visual scene to the crew (image) allows pilots to see the external environment: land, infrastructure, day, night, rain, fog and clouds.

The equipment in the cockpit is strictly that of an airliner to the point where the equipment is interchangeable with those of an airplane.

Behind the cockpit part is the instructor station with controls allowing to progress according to the scenario of the session. The Instructor's station is positioned directly behind and between the two crew members. This facilitates intervention and teaching when required by the students (Fig. 1).

Apart from cost savings and adverse environmental impact (noise, CO_2), simulators have the ability to create complex situations which would be impossible and unsafe to recreate in a real airplane.

For example, invading the cockpit with a thick smoke that can be from electrical or pneumatic origin. Or simulating the failure of 2 jet engines on one side as well as the simultaneous extinction of the 4 jet engines (due to volcanic ashes for example). Another example would be an exercise flying with flight controls jammed.

The simulator can also recreate environmental situations such as severe windshear which in reality could lead to ground impact, severe turbulence, dramatically reduced visibility, icing, heavy rains or strong crosswind.

The simulator also allows for intervention with other traffic, whether it is simply to enhance the realism of the experience, or to allow the pilot to practise procedures for avoiding imminent collisions with traffic on the ground and in flight.

Yet the simulator also provides the instructor with specific opportunities to enhance the learning experience. They are:

- The position freezing: the aircraft continues to "live" normally. His handling is necessary but it remains "suspended" in space, vertically from a given point.
- The total freezing which allows to freeze the situation, e.g. during the time of an explanation.
- The repositioning which reverts to a position chosen by the instructor in the airplane configuration desired. For example on final, runway track, landing gear and flaps extended with one engine inoperative.

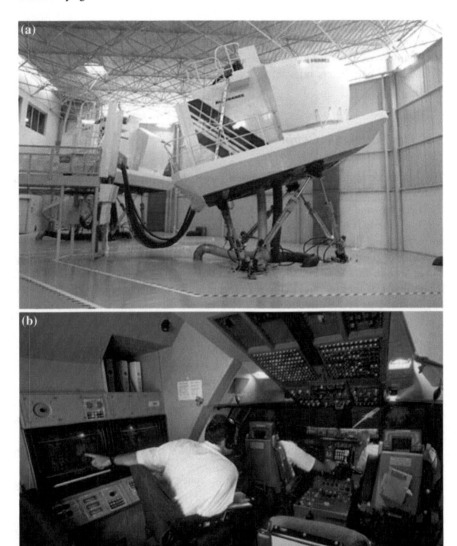

Fig. 1 Example exterior (**a**) and interior (**b**) of a full flight simulator

- Some parameters about the aircraft such as gross weight, center or fuel can be changed at will.
- Finally, one can subsequently observe deviations from the optimal trajectory and other parameters relevant to the performance of the pilot.

1.2 The Simulator Cabin

This simulator represents a generic aircraft. It is used primarily to work mainly on global crew behavior, in case of emergency.

The pilots and cabin crew train together once a year on this simulator.

This training is also very useful but to adequately cover its interest would require an independent chapter.

So in the present book we will limit ourselves to the presentation of the training and control in simulated situations for pilots already qualified on a type of machine, in the fields of their annual skills maintenance and of their training related to the operational changes and feedback "REX".

2 Instructors and Pilots Involved in the Simulated Situation

Let us precise that we work here in a very constant regulatory framework that is now European. These are the Joint Aviation Requirements.

Each pilot must renew her/his license annually on the type of aircraft on which s/he flies. This is called skills maintenance.

Thus, an Air France pilot, has to perform each year 2 trainings and 2 checking sessions in the simulator.

Each of these checks, including the on line check during a real commercial flight, can be failed. In such case the pilot will have to perform remedial training followed by a re-check before being allowed to fly again.

For a standard simulator session the crew consists of a Captain and a First Officer. They will be alternatively "Pilot Flying" (PF) and as the "Pilot Not Flying" (PNF).

PF is who controls the trajectory of the aircraft using hand controls or autopilot. PNF is who supports the other pilot by carrying out other tasks, such as radio. The whole work is performed according to a standard task sharing, with many "cross checks". Although both pilots may have the same overall level of knowledge conduct these aspects and the technical management of the aircraft, the commander is the decision maker on board, and if time permits, different options have been proposed by other crew members.

The instructor or controller (see Appendix) conducts the simulator, observes and notes the significant facts, the operation of the crew, for debriefing and scoring.

He also simulates the other radio messages from the ground (air traffic control, company) and cabin crew from the cabin.

Insert 1
The instructor and the examinator.

The instructor may, depending on the planned actions, be a captain or a first officer qualified on the aircraft type.
The pilot's license must be valid for certain types of training.
S/He acts on behalf of the company or training center.

The examiner has to be a captain holding a licence and rating granting privileges at least equal to the licence/rating for which they are authorised to proficiency checks.
The pilot's license must be valid.
S/He acts on behalf of the aviation administration, the Directorate General of Civil Aviation, who applies the standards of the European Aviation Safety Agency.

In order to maintain the validity of their qualifications as instructor or examiner, individuals must practice a minimum number of annual of training sessions, attend specific conferences and have to be checked while training or controlling every three years.

3 Structure and Constraints in a Simulated Situation

Within skills maintenance sessions, a simulator session consists of a 1 h briefing, 3.30 h simulator session followed by a 1 h debriefing (Fauquet-Alekhine and Labrucherie 2012).

The simulators are operated twenty-four h a day, seven days a week, and crew session turn-around times can be as short as 5 min.

Punctuality, time management with a strong focus on the session objectives in the simulator is thus absolutely necessary.

However, the sessions are intense in order to meet the regulatory requirements.

They include some mandatory requirements, set by aviation administration to renew the license.

Additionally, the company will introduce a few scenarios where potential risks have emerged through the analysis of flight data, confidential pilots' feedback, international statistics or recent events.

The skills maintenance training is organized to review the necessary skills required in various subjects, on a 3 year cycle.

Insert 2
Example of a training program.

The beginning is a night training at Mexico meaning on a high altitude, surrounded by obstacles.
Cockpit preparation and engine start.
Take Off experiencing strong crosswinds conditions.
Emergency operation due to ground proximity.
Manual approach and landing, without flight director, with strong crosswinds.
Takeoff with engine fire. Engine shutdown and fire extinguish.
Return to airfield with engine failure.
Go around due to loss of visibility.
Night Visual circuit and landing with an engine failure.
Strong crosswind takeoff.
Failure of both FMC (Flight Management Computers) involving great loss of navigation and performance information.
Approach and landing under these conditions with strong crosswinds.
Training continues at San Francisco.
Severe engine damage during take-off.
Return with engine failure and side step on final for landing on the parallel runway.
Take-off including thrust adjustment due to a wind shear.
Return to airfield with turbulence during landing due to wind.
Very low visibility during take-off.
Approach with fog and navigation instruments failure during landing phase.
Go around.
Return to complete stop with a visibility of 125 m and DH 17 feet or 5 m.
Taxi to parking stand.

Of course, during these simulators sessions, risk management must be demonstrated by the crew. This is done using a classification by types of accidents, as shown in Fig. 2.

We immediately can observe that "Controlled Flight Into Terrain" (CFIT) and "Loss of Control in Flight" are particularly fatal.

Analysis shows despite a large number of landing accidents (15), the number of victims is relatively low (136).

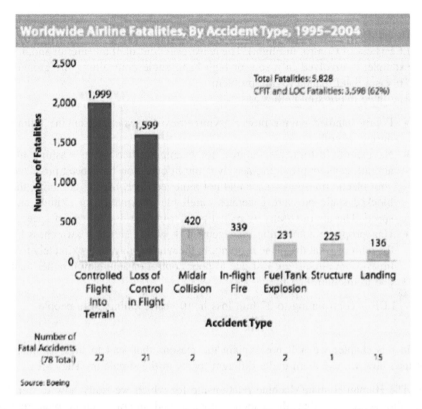

Fig. 2 This figure shows that over a period of ten years, 78 accidents occurred in commercial aviation worldwide, killing 5828 people (Boeing datas)

Insert 3
CFIT (Controlled Flight Into Terrain):
The plane is perfectly under control but hits the ground. This is a navigation problem. The two main subfamilies of CFIT are:

- Collisions with the ground during management of a failure. The crew is using all his resources for the management of the failure and hits the obstacles or the ground.
- The non precision approaches which are approaches without guidance plan for the final segment.

CIFT is alone responsible for the crash of 28 planes in 10 years, killing 1999 people.

Insert 4

LCF (Loss of Control in Flight). The crew loses control of the aircraft and, for example, is involved in a spin or high bank angle until hitting the ground. This is a hand manual flying problem.

The three major subfamilies are:

- Poorly handled engine power asymmetries: for example, engine failure during takeoff.
- Navigation Instruments failures: for example when while washing the aircraft, sensors providing aerodynamic information have been protected with plactic transparent tape and not removed after. During climbing, the blocked static pressure generates unreliable information of altitude and speed. The aircraft stalls by reduction of wing lift due to loss of speed.
- Unappropriate automation management or wrong situation awareness by the pilot of what the plane is doing. In the world today, approximately five times more aircraft are lost due to poor control of automation rather than due to uncontrolled fire.

LCF is contributing to 27 hull loss in 10 years, killing 1499 people.

In this chapter we will not explain the reasons that lead to these results, but instead how we use them under different topics in the simulator. They are:

- The Human/Human/Machine relationship for which we study how to run this "threesome" so that the captain, copilot and aircraft action plan remain consistent.
- The relationship to the environment that greatly contributes to the "Controlled Flight Into Terrain (CFIT)".
- The relationship to time since the error of the front line operator in a dynamic situation represents 80 % of the root causes of accidents.

Given what we have seen above, we understand that strict organization of work is needed to deal with the abnormal situation.

It results in a "golden rules" approach that we will see later.

Given the number of scenarios during the session, we understand that it is impossible to manage till completion for all of the planned scenarios.

Until recently, each instructor was doing his best to finish in the allocated time, stopped the exercise during the sequence, froze the aircraft to resolve a situation to avoid being too far from the airfield.

All this was affecting the standardisation of training, distorting and especially for the crew, sometimes hiding some major risks such as the CFIT. For example, in Lyon, safety altitude is evolving from 3000 to 4000 feet (900–1200 m) at 17 nautical miles (32 km) southeast of the airport, i.e. a flight time of 4 min.

To avoid losing time to return, a few instructors in the simulator freeze position abeam of the runway, the time to manage the incident, while ignoring whether the crew in the plane flying at 3000 feet would have received the CFIT risk (subfamily collision with the ground when managing abnormal situation).

Thus, to be more homogeneous and better learn how to manage risk, we create three classes of the main training categories.

Class 1   the crew used the aircraft optimally. We study the relationship Man Man Machine without additional consideration.

Class 2   where class 1 is performed in a potential hazards environment (Ex: CFIT), under time pressure.

Class 3   the situation offers in addition to class 2 a complex environment, economic and commercial implications, with management of passengers and cabin crew.

For effective teaching, it becomes necessary to define level of intervention by the instructor and knowledge of the scenario by trainees, for each class.

Insert 5

Classes of Training Purpose
Class 1:

Relationship man/man/machine apart from all environment considerations. This is typically the learning process or review of procedures and know-how. Either manual flying or automatic flying can be required for training purpose

All the tricks of the simulator (repositioning, Freeze position, etc.) can be used.

For emergency procedures, the abnormal situation management with the golden rules is followed by applying it only to the black writing.

The crew knows the scenario and the instructor intervenes when necessary.

Class 2:

Relationship man/man/machine during a local flight at a given airport, or on a particular route, with CFIT risk.

The simulator tricks that destroy the dynamic situation are prohibited in order that the crew experiences a context of CFIT risk, under time pressure.

The crew operation is studied in terms of resources, workload allocation, task sharing, situational awareness, etc.

The "short term" decision should be trained.

Ex1: Captain Decision for task sharing then go back to the scenario with PF designated by the instructor.

Ex2: Elaborate situations involving safe decisions from captain or first officer (unstabilized final approach, landing interruption due to loss of visibility, etc.).

The operations limitations are strictly observed and any excess, for example crosswind presence, requests a decision to go around.

The golden rules guideline is partially applied (written black and blue part including "short-term tactical decision" aspect).

The crew knows the scenario, however, it must allow a few surprises or doubts and provide "gateways" unusual for a given procedure.

Example: go around on approach by very low visibility due to engine failure rather than a classic engine failure after go around.

The instructor intervenes when needed.

Class 3:

It is the Line Oriented Flight Training (LOFT) with different realistic options, a few pauses, as close as possible to the real flight.

The decision making often require a balance between internal risk (related to the crew expertise) and external risk (related to the situation) and may jeopardize the continuation of the flight (difficult decisions, possibly medium term decisions).

The decision making process is the main training objective of LOFT.

The whole Golden Rules guidelines must be used.

The least known scenario possible.

Instructor intervention is strictly prohibited...

We have to define the targets to be achieved by the "golden rules".

Insert 6

Golden Rules guidelines
(Class 1: make the black part.
Class 2: make the black + blue
Class 3: Perform the entire guideline)

Failure annoncement (the first who is aware of it).

FLY/NAV (This is the first priority because CFIT and LCF).

PROCEDURE.
memory items performed and cross checked by the pilot not flying).
Ex: the case of fire ENG 4 during take-off.
Landing Gear Retracted
ENG 4 iddle
ENG 4 master OFF
ENG 4 Fire push button pulled
If ENG 4 light still firing, first fire extinguisher Dish
After 30 seconds, if ENG 4 still firing, second fire extinguisher Dish

All these actions are performed with cross checking procedure to prevent panel control
or engine mistakes.

At this stage, the aircraft is usually under a safe controlled situation
(If the fire continues, one must evolve to the emergency landing, if necessary
overweight, meaning a gross weight greater than the maximum landing weight)
... ...

Here, the captain decides on the task sharing
The PF manages flight control/communication: fly, navigation, radio communication,
remaining fuel.
With flying and navigation, he faces the LCF and CFIT risk.
It is therefore essential that he dedicates much of his cognitive resources (teaching 80% for
imaging) to "fly the plane." The rest (20%) is mainly used for cross checking and maintaining
the mutual synergy crew

The PNF manages the failure in terms of understanding the situation and runs checklists. His
main resources are oriented to these tasks. It is understandable that in a real failure or a
LOFT, the captain keeps for himself this management which can be very complex. It is to be
also noted that in three quarters of the accidents, the captain was in charge and did not have
enough resources to manage both the failure and the flight of the aircraft.

MANAGEMENT FAILURE   split 80/20   FLY/NAV/COM/FUEL

ABNORMAL CHECK LIST
                                                                                CFIT / LCF Risks

CL NORMAL
(After takeoff
Descent / Approach / Landing)

PAC
(Abnormal Procedures Supplementary)

PNUP
(Normal Procedures specific use)

STATUS   Technical
                Operational
                Commercial
                Economic

DECISION

INFORMATION   (Control / Crew / Passengers / Company)
REX   (Reports / FeedBack possibly anonymous)

Note: It is very important that trainees know in which class they work.

The syllabus and objectives must be perfectly clear.

In this way, the rule of the game is known for trainees and instructors and ensures greater standardisation of training and checking of our 4000 pilot, trained by 450 instructors.

3.1 The Desired Targets

If we consider that the job of a pilot is generally to fly from point A to point B:

- The safest as possible
- The most punctual as possible
- The most economical as possible
- With the best customer satisfaction as possible.

And if we consider that to reconcile these 4 "possible" must mean constantly to adjust, to choose with pragmatism and common sense to make the flight as safe as possible and therefore:

- to know the risks
- to manage threats and errors that lead to them.

Never to forget that 80 % of the root causes of accidents are caused by crew errors.

It is in Class 3, or LOFT, that we try to create a real flight. The instructor simulates the presence of the purser, and therefore passengers' problems.

However, classes 1 and 2 remain quite necessary for the training with or without the time constraint, as we have seen above.

Let us suppose we are flying from Boston to New York. It is a pretty nice weather in Boston and New York. The captain pilot (PF) is in charge and the first officer is assisting (PNF).

During takeoff, we experience an engine failure number 4 when the pilot rotates the aircraft to leave the ground.

In class 1, it is checked:

- That the captain controls the asymmetry engine power flight (see LCF risk).
- That the followed trajectory is perfectly the forecasted one in case of engine failure (risk CFIT)
- That the captain announces the title of the appropriate procedure, here "Procedure ENGINE 4 FAIL".
- That corrective actions are performed in mutual control by the crew.
- That the actions of the checklist box (memory items) are checked by reading again (PNF).
- That the emergency and normal after takeoff checklists are performed.

If performance is not at the required level, the instructor can reposition the airplane at the take off for a similar exercise.

In class 2, we check:

The same as in Class 1, but this time the crew must handle all the problems by never freezing the flight of the plane.

However, the captain may ask the copilot to monitor the flight controls and navigation in order to better focus on the abnormal situation management and on the compliance with the golden rules as discussed above.

Only parties typed in black and blue of the golden rules have to be performed. Indeed, here in class 2, the scenario requires the destination for New York and for the captain, the question of returning to Boston is irrelevant.

If the captain, for managing the failure, has delegated "Fly the plane" to the copilot, the instructor intervenes when necessary to ask the captain to be PF again.

In class 3 (LOFT):

The intervention of the instructor is strictly prohibited. The crew handles the problem as he sees fit.

The captain must consider the weather in New York and perhaps decide to return to Boston where it is better.

He will have to coordinate with the purser, make an announcement to the passengers, and handle all problems with the company.

In 75 % of accidents, it is the captain who is PF. Indeed, in this case, busy by flying the plane, he has few mental resources to analyze and decide. In addition, in case of captain poor performance, the co-pilot take over remains tricky, even if he is trained for. For these reasons, the opposite task sharing is recommended and it is likely that the captain gives the controls to the copilot for the rest of the flight.

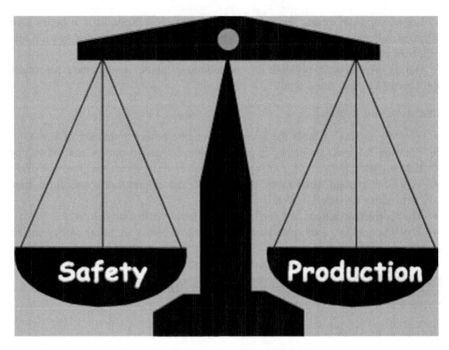

Fig. 3 Operational behaviors—achieving the system objectives. "One can argue that the mark of the expert does not lie in his years of experience and practice of aviation, but rather in how well he masters the skills necessary to manage the tradeoff between production and security" (extract from LOSA)

So, it is obvious that the LOFT is the closest to real flight. However, the situation of simulation is not representative of the balance safety/production that must be managed in real flight (See LOSA: Line Observation Safety Audit on which we study the actual flight crew operations, particularly threats and errors management, see Insert 7) (Fig. 3).

Indeed, in the simulator, the decision to return to the departure airport is psychologically easier, because even if we pretend, passengers are not really there.

Safety is the main concern for the industry, and so it is for the instructor. In the simulator, people will naturally be safe, whereas in a real flight it is different.

And for a non pilot reader to understand it well, let us take an example of my personal experience: a flight from Pointe-à-Pitre to Paris.

About an hour after take-off from Pointe-à-Pitre to Paris, the chief purser tells me that a passenger has faint.

The initial diagnosis made by the crew leads to a heart attack. After a call on the public address, luckily a cardiologist shows up and proposes to evaluate the situation within 30 min.

I went back to my cockpit and began to consider different alternatives, depending on the critical status of the passenger. Back to Pointe-à-Pitre? Diversion to Bermuda, but would she be treatable if her financial guarantees are insufficient? Diversion to Santa Maria in the Azores, where medical resources are limited? Diversion in Lisbon, not too far from the flight plan but 5 h later? Continue to Paris which is six and a half hour flight? As agreed, after 30 min, the doctor came in the cockpit to tell me about his diagnosis. To my question about the urgency, the cardiologist told me: "From now on captain, every hour gained will be valuable for the life of this woman." This answer did not satisfy me and I enounced the possibilities with their restrictions. Of course the solution to Paris was by far my best operational option. As an answer to my 6 h and a half flight time question, he replied: "I can only repeat myself captain. Every hour saved is valuable for the life of this woman". The first officer and flight engineer (it was on the B747-200, with flight engineer) did not seem eager to give their opinion. But something had to be decided. Once the doubtful options identified, I could return to Pointe-à-Pitre with a 3 more flight hours, 40 tons of fuel to be dumped to reach the maximum landing weight, a 1 h 30 stopover and therefore a 5 h delay to Paris. Obviously, it will generate higher cost and a big trouble for other passengers. We could also continue to Paris, but jeopardizing the lady's life, a thing I could never forgive myself.

Finally we returned to Pointe-à-Pitre.

Can we simulate such a scenario? Of course not, because even a loft with this type of case, "sick passenger," are only words. I cannot believe having such a loneliness feeling I had facing the decision that day, during a simulated session.

But these issues showed on the desired targets being asked, let's move on to the simulator.

3.2 Description of the Simulation Session

3.2.1 The Briefing

During the 1 h briefing, in the past, the instructor used to introduce the syllabus of the session by presenting some technical reminders and application of procedures.

Due to the number of subjects needed to be highlighted, the briefing was too long. One way communication was observed with no space for interactivity.

Also, to further involve our trainees, we have introduced the briefing by questioning.

For this, before the start of training season, we send both pilots and instructors a collection of briefings to be discussed during the sessions. This booklet is the same for instructors and trainees, and each one has the same level of information.

Related subjects at a given training sessions shall be prepared by the trainees to review the knowledge and key points, but also by the instructor to ask questions.

These questions should be used to verify that trainees know the basics but also to explore some fuzzy points or nice to be known.

We do not want to implement standard questionnaire, at first because it would be quickly known and secondly we do not want this quiz to be a test. This is more an interactive discussion among colleagues, in which the instructor must proceed with tact and avoid pushing his trainees to the corner before entering in the simulator.

It is rather for him to create a relationship of trust with its trainees.

If serious issues are found, they will be solved by the instructor before the session and then again during the debriefing.

To avoid pitfalls, we have trained instructors to this new method and provided a DVD that allows them on real situations to review various aspects of the briefing by questioning. In addition, instructors are regularly supervised in training situation during recurrent training at simulator by senior instructors. These checks are also part of the required training tools requested by the administration.

After a year of experimentation, this method satisfies the whole people.

Indeed, trainees who have well prepared their briefing easily demonstrate their knowledge of basics and key points. In addition, discussions between colleagues on sensitive points allow them to give their point of view and also improve the overall knowledge. Finally, it avoids the former long briefing trying to cover all the necessary knowledge. This was intended to cover the shortcomings of the weakest but boring for those who knew well the subject. Depending on the time of the session, sometimes at 5 o clock in the morning or at night, this could be an ordeal if not difficult, at least not motivating.

Here, with a well-conducted briefing by questioning and the interactivity it provides, the instructor can quickly intervene at the required level, based on answers but also on questions from the trainees.

But remember this key point: at the briefing, the instructor must have successfully created a relationship of trust between him and trainees to get the most out of what will follow.

3.2.2 In the Simulator

The recurrent training includes four sessions of 3 h 30 min each.

Two of these sessions are training sessions, the two others are checks and can lead to a failure and not permit the trainee to fly.

Sessions and E1/C1 E2/C2 are spaced approximately out of 6 months.

The architecture of the sessions on the one hand, the level of intervention by the instructor on the other hand, may be very different from one session to the other. In this sense, the pedagogic "classes" we've seen before allow to better establish the rules of the game and to adjust the distance instructor/students according to objectives.

Once in the simulator, the crew moves while the instructor initializes the simulator.

Then the lesson plan runs automatically in sequence, but allows the instructor to select particular options of the scenario.

For example, for an engine failure, he can choose the engine that will fail to ensure that the pilot properly controls the asymmetry of thrust. For this, the rudder should be immediately oriented on the right side, with the right amount. This is done on foot, through the pedals, and side error causes the crash almost immediately. The engines power asymmetries poorly controlled are one third of loss of control in flight.

The instructor can also decide when the engine failure occurs.

S/He can leave if necessary the auto lesson plan to trigger events desired by her/him. The purpose of this is to avoid the trainee to anticipate the scenario of the session because it would be difficult to assess her/his skills.

But in any event, the overall scenario will be respected, both for educational reasons and for checking, but also on the regulation point of view.

Throughout the session, the instructor has several roles:

- operates the simulator.
- simulates air traffic control radio communication and potential contact with cabin crew or ground staff.
- ensures the level of intervention in accordance with the "class".
- observes the technical and human performance of the crew.
- must take notes in order to design the debriefing and then carry out assessment of trainees.

We see that the instructor workload is very high and that in order to ensure properly the five points above, he must have well prepared his session first.

Overall, for an experienced instructor, the first two points are not a problem. But the introduction of "classes", for some of them, remains still a problem for an adequate level of intervention, depending on the class.

For example, LOFT (Class 3) a few teachers are struggling not to intervene if the crew are going totally in an unexpected direction in the alternative scenario. They fear that the time does not allow the flight to proceed the whole flight till landing. Yet this does not matter, since the main objective of LOFT is the study of decision-making. On the contrary, it would be very interesting to analyze during the debriefing why this crew adopted an unexpected solution, which can sometimes be very effective.

This is better than handing the crew "back on track" and watch a landing, which often has little interest, even in a degraded mode.

But an instructor, although held by the time constraint, likes to "finish the session" and for that reason, sometimes falls into the shortcomings of the intervention or the use of not necessary shortcuts in the simulator to the current class.

Observation and notes taking of the technical aspects are not a problem for experienced instructors. Their knowledge in this area is at a very high level.

But it is very different in the working of the individual and professional group.

However, it was fairly easy to set up sessions CRM (Crew Resource Management) major concepts concerning human factors:

- The individual: memory access, acquisition and information processing, situational awareness, action plan, etc.
- The professional group: communication, leadership, cooperation, decision making, etc.

Yet it is difficult to train instructors to observations of those subjects in the simulators. Why such a difference?

First, because the CRM sessions take place in classrooms, the observation is based on media movies designed for the course, depending on the chosen theme. The two facilitators are fully trained on the product and can easily help the analysis and propose areas for improvement.

In the simulator, the strengths and weaknesses are different from one crew to another, or even a trainee to another. The instructor here, must discover them.

On the other hand, as for technical aspects, the requirements to review all the human factors issues are based on a three-year cycle.

As it is difficult to observe everything at the same time, we chose to look at a topic in details on a chosen session. The others are observed more superficially, and if necessary would be reviewed. Attached are three examples of Grid Observation Simulator (GOS) respectively focused on the decision, the situational awareness and the stress. When comparing, you understand the relative importance given to the chosen theme.

The reader will find also two extracts of NOTECHS (Non Technical Skills) showing effective and ineffective practices in terms of crew performance for the concepts studied and the negative effects of stress.

Insert 7

NOTECHS and LOSA

After several years of practice of CRM in airline companies, the scientific community has observed that while undeniably a CRM culture was born, it was difficult to assess the gains in performance and crew safety.

NOTECHS was the answer given by a group of European experts.

They described poor or good practice for concepts such as decision, situational awareness, leadership, cooperation.

(See Appendix)

LOSA or Line Observation Safety Audit was the North American answer by observing, in real flight exclusively, threat management and crew errors.

Obviously, a not well prepared instructor will have difficulties to make the necessary observations.

It is absolutely essential that before the start of the training season, s/he looks at what moment, in the scenario, he has a chance to observe the crew performance about the concept which has to be studied on the session.

The Human Factors training department provides help for these few points of observation and for advice to deal with the debriefing. But this does not exempt them completely from a major work to be effective for both the observation and to conduct the debriefing. Indeed, they are responsible for making the link between performance and technical reasons for this performance.

Let us consider the stress concept studied on the training mentioned at the beginning.

During all sessions, there are several engine failures and, according the scenario, the captain has to demonstrate his skills as PF with the first officer as PNF, then, on the other engine failure the reverse role.

It is a very interesting symmetrical situation to observe the stress of each one and then to conduct an effective debriefing.

Indeed, stress is a very personal phenomenon, and not easy to observe, especially with people used to hide it, because it had a very negative connotation for too many years. They preferred to hide or ignore the problem rather than learning to manage it.

3.2.3 The Debriefing

At the end of the simulator session, the crew gave everything. They are tired.

A one-h debriefing was planned at the end of the session.

Especially if the session is a check, the instructor must give the result that most of the time is positive, with license renewal. Indeed, the crew should first be reassured, for a better listening later.

We will say a little further a few words about the difficult case of dealing with a failure.

The overall result is indicated, the instructor leads the classical technical debriefing that follows roughly the chronological order of the session scenario. It is an effective guideline for students and convenient for the instructor, in line with his notes.

During the technical debriefing, the instructor provides guidance for correcting weaknesses and reinforcing strengths.

On the other hand, he puts on the blackboard some consensuals significant facts, used after to analyze human factors.

This technical debriefing lasts about 30 min and at the end, each student can get a pretty good idea of the instructor grading that would be reported on the check forms.

The trust is again a key element here, since this step being achieved, the instructor will propose to start with the human factors analysis of the elements noted on the blackboard. It is easy to understand that for students to reveal themselves, they must feel very confident that nothing they say will be held against them, and with no negative influence on the instructor's rating.

The instructor is used to questioning the first officer first, to avoid the superior's influence by the captain.

S/He behaves here as a facilitator to guide the analysis and generate a lot of interaction and exchanges, especially between themselves. Then, we can observe the emergence of an invisible part, hidden till now in everyone's mind. This permits to understand a lot of things and provides a better awareness.

Sometimes this can lead to psychodrama as I experienced on Boeing 747 fleet at the beginning of these experimentations.

A particularly autocratic captain had made life difficult for his co-pilot and flight engineer for the duration of the session, annihilating any suggestion with the back of a hand, using an abrupt tone.

During the debriefing, respecting the golden rule, I questioned the co-pilot first, and asked him why, when the captain did not consider a particularly relevant suggestion, he had not insisted, while this is recommended by the assertiveness models.

This was a great silence, you need to let it last long, because silence is an efficient tool to express a meaning. In order to break the silence the captain told me: "I feel that I scare them."

I replied that he should not ask to me, but to them. He did so, generating a huge amount of criticism from the flight engineer and the copilot. They described him as an awful captain. He was shocked and asked me again: "Am I like that"?

The beginning of my answer was "From my seat, a little bit but...."

Then began a long process of analyzing and understanding the reason why, and how. No one wanted to leave the room before 2 h of explanations, going far beyond the official time of debriefing.

Finally, they made peace, but after the flight engineer and the copilot have left, the captain felt the need to stay an extra hour to ask me some advice about how to implement the relevant leadership style.

This trial period has led to the establishment of a method to have all of our 450 instructors able to:

- Observe as we have seen above.
- Then debrief the Human Factors aspects in three distinct steps:
- The extraction of some significant facts written on the board.
- The interactive analysis.
- The establishment of improvement axis for everyone.
- The ability of our instructors to realize this is a major issue.

We must dig into this 80 % of accidents root causes, which are due to errors or inadequate management of the situation by the crew.

This is the reason why we train our instructors in the classroom and provide them a customized support for the simulator session. We have also established a Human Factors website dedicated to aircrew, allowing them online training. We then check their performance during the simulator sessions.

Let us go back to our stress example, difficult to observe and sensitive for the debriefing conduct.

During the observation of an engine failure, the instructor has certainly found a number of technical facts, but can s/he be sure of his observations on stress, if s/he has been able to make any?

The best way is to ask each trainee to point out where they think they position themselves on the stress/performance inverted U curve graph, when operating as PF and PNF during the abnormals situations.

Performance

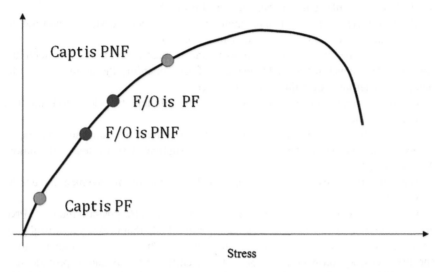

Fig. 4 Inverted U-curve drawing arousal and performance comment: performance will improve at first as arousal increases, beyond the optimal level of arousal the performance starts to deteriorate if too intense stress is observed (breaking point or «stall»). Some typical signs of excess stress will be observed. Adapted from Nixon (1976), Yerkes and Dodson (1908)

For this, the instructor draws a blank inverted U-curve on the blackboard and asks the co-pilot to plot on it her/his two operating points. Then, the captain does the same (Fig. 4).

Insert 8
Wing lift curve
Aerodynamic lift depends on several factors including the angle of attack of the air over the wing.
This is represented by a graph, characteristic of a given wing.
In the area of low angles of attack, lift increases with angle. But when the angle becomes too high, the air flow becomes disturbed, vortex appears. Lift collapses and the plane swing forward. This is the stall.
Hence the analogy of the curve stress, especially meaningfull for the pilots.

The instructor asks the co-pilot to plot on the curve her/his two operating points, then captain does same.

Then, it is very easy to make them talk. At first, about their feelings which lead them to this plotting. What symptoms of stress: physiological, mental block, tunnel vision or other sign?

The instructor will show then why facing the same real engine failure risk, the level of stress is different whether they are PF or PNF.

Then we review the stress management process in which individuals interpret the situation they have to face versus their real know-how.

It will also be interesting to compare the captain's point of view with the copilot's. This can bring to light negative effects regarding the confidence in the other, or sometimes regarding self-confidence.

It is finally important to know how they managed their stress level to bring it to acceptable values.

By simplifying the problem? Decrease the workload by using automation, or otherwise disconnecting them if they were disruptive? Use of sense of humor? Another way?

When this interactive analysis is completed, the areas of improvement have to be established.

In the case when the co-pilot is more comfortable in a PF position than in a PNF one, we should consider her/him as a «hand skill gifted» that is to say a skillful pilot who flies sharp in asymmetric flight. But, perhaps, s/he is a bit too much relaxed and the necessary knowledge to conduct efficiently checklists and to provide support to the captain could be lacking to him.

In this case, the guideline to progress will focus on knowledge and reading some benchmarks to improve a supportive attitude.

If instead s/he is more stressed in PF position, s/he is probably a pilot less comfortable with the manual flying, but compensates with a hard working side that makes her/him comfortable in her/his PNF tasks.

In her/his case, the guideline to progress will be to review certain basic manual skills in engine power asymmetric flight situations, although safety has not been at stake, as s/he passed the check, to propose an additional training session for polishing.

On this non-graded session, the instructor will have her/him work on specific exercises such as the coordination with the engine thrust and the rudder to counteract the engine power asymmetry and other various settings.

In any event, if they are successful with their simulator check, they must leave feeling confident with their ability to fly their planes tomorrow safely in line with the standards of civil aviation majors. That should not prevent them from feeling the necessity for improving.

Still remaining is the sensitive case of dealing with a simulator check failure.

I will say a few words, but here too, the result must be given early.

Then, in full respect of the person, it is necessary to understand the reasons of the failure to suggest effective proposals for retraining.

It is sometimes necessary to provide a break or flight planning arrangement, time to recover from an "accident of life: death of a loved one, separation.

In some cases, medical or psychological assistance will be recommended.

As for the recurrent training, it is quite easy to be positive, because downtime and retraining will have no consequences for the career path.

It is much more complicated when it is a failure that prohibits the entry into the company or an upgrade as captain or even a change of aircraft type.

The personal, family, professional consequences are very important and require thus support, again in full respect of the person.

3.2.4 Grading

Fourth and final part of the session, the grading reported on a document shows the performance of every pilot, whether for training or checking. Only in the last case, the examinator is required to make a proposal of pass or fail which will then be endorsed by the Head of training of the fleet concerned.

The graded main categories are: manual flying, knowledge, professional behavior and commercial aspects.

These sheets are registered in his professional records. But with 4 sheets for the simulator, one more sheet for the line check, 5 sheets per pilot per year, it leads to over 20,000 sheets per year for some pilots managers in charge of professional standards.

How can they manage so many informations, and on what purposes?

Two years ago, we went from paper to computer.

Always looking for improvement, but also because the European regulation is changing, we are currently working on a new rating sheet, used since 2008.

This support, very innovative, illustrates the skills required to do this job.

The grading for each skill varies from 1 to 4.

Grade 1   represents unacceptable performance impacting flight safety and leading to failure.

Grade 2   is acceptable but improvable.

Grade 3   is the standard.

Grade 4   represents superior performance, showing that practices should be brought to the attention of all and can sometimes generate an improvement in procedure or practice.

In case of grade 1 or 2, sub-categories appear to explain the reason of a performance below the standards.

The two sub-categories are knowledge, and human factors by using the criteria NOTECHS.

In fact, this electronic grading has several goals:

• For the trainee: to allow a feedback on her/his performance during the session and to give guidance for improvement.
  This grading fills in her/his professional file (e-file). Warnings prevent any loss of performance, for example related to an accident of life or easy going behaviour.
• For the Instructor: s/he may position her/himself in the pool of instructors. Is s/he relaxed or tough? Is s/he maniac always grading 2 the same categories?

In short, s/he will have a tool for a self evaluating process.

Her/His superiors will assess if s/he deviated, and may take the necessary actions to help her/him correcting.

From a systemic point of view, it will be possible to analyze fleet by fleet strengths and weaknesses of the pilots.

Example: on this aircraft type, long landings can be observed (measurable in the simulator and on flight by flight analysis).

The sub-categories: human factors and knowledge will provide some insight about what and how we must act to fix the problem.

Besides grading, the instructor can feed the system with a "REX".

Indeed, the REX, in all his options (systematic analysis of flight recorders, reports, verbal or anonymous REX) provides information on the precursors or incidents relating to all in line flights. But most flights do not face likely degraded or complex situations offered by the simulator.

In fact, all the emergency procedures are very scarcely observed during real flights and the best way to assess crew performance before an emergency is the simulated situation.

In case of recurring problems, the REX instructor can determine the root cause. If it is a training problem, a short process within the training department should be used.

But, if the cause is a poorly designed or inappropriate procedure, leading in some cases to a voluntary deviation for a better efficient situation management, a long process involving the main departments to improve the procedure should be used. This sometimes requires the agreement of the aircraft manufacturer or of the administration.

4   Elements of Conclusion

It is clearly understood why the simulated situation is a tool for a pilot and a pilot group to learn, grow, creating trust.

This last point is particularly important and generates a discussion among trainers.

The human factors study shows that an operator feels comfortable, confident, when running at 80 % of its peak performance.

In addition, when approaching 100 %, its confidence deteriorates (Rizzo et al. 1987).

Then a question naturally arises: should we go beyond the limits in simulated situation to cope better with real flight?

For example, the crosswind limit the B747-400 being 30 knots maximum, would it be wise to train with 40 knots in the simulator to increase know-how and to create confidence in the flying skills mentioned above?

Detractors fear that such a confidence leads to exceed the limits in real flight.

But, we might think that speaking to professional pilots with training objectives clearly defined, the limits would be strictly applied.

For example, it might be acceptable to exceed the limits for Class 1, as it is typically in this class that such skills are acquired. In class 2 and 3, any over-limit would be prohibited and on the opposite, the safety decisions would be highlighted. Thus, the announcement before landing of a crosswind exceeding 30 knots, in Class 2 or 3 should, except in emergency situations, lead to a decision to go around.

Another issue is also under discussion.

In case of loss of control, should the crew being let, up to the accident or is it better to freeze the simulator before the crash occurs?

The people for the second option want to avoid the crew to be shocked, and experience a potential loss of confidence.

But, on one side, errors occurring very close to the ground, will not allow the instructor enough time to freeze the simulator before the crash. On the other hand, in some situations, going to the end, some crew develops creative survival strategies while others, thinking that it's over, are more likely to stop fighting for survival.

Finally, it seems reasonable to assume that aviation professionals must be sufficiently strong on the mental aspect to face their own death, even simulated.

For these three reasons, I am personally against the majority. I think it is better to go "to the end" but to know how to demystify the event, to draw out of this the lessons learnt and to propose an equivalent situation to verify this time that we reach a happy end.

Finally, let us see how the simulated situation sustains the "great system".

As the REX from the line, the observations in simulated situation led us over time to change our attitude regarding the error concept.

The man, intelligent but limited, agreed to be weak to perform.

"Correcting performance and error are Two Sides Of The Same coin, we can understand this: the error is related to human intelligence as two sides of a coin" (Reason 2008).

The error has become acceptable, the same as it is to live in a world without error appears unsuitable as would be a desire to raise a child in a totaly sterile environment. He would not build defenses and the first virus would kill him.

Through a series of defenses (Reason's model), aviation has organized a system in which the error or undesirable event are not serious if blocked by the defenses, they have no consequences. However, thanks to REX and observations in the simulator, the error is closely monitored to attempt to produce fewer, but above all to constantly improve the defenses.

But other major changes are popping up.

Dr. Rene Amalberti brilliantly described in his article on the "ultrasécurité" (Amalberti 2001), the issues our industry faces nowadays.

Rose quickly through the three stages of the life cycle of socio-technical systems (heroic times, time of hope, and time of justification), aviation has been elevated to the rank of the industries in which the accident has become unacceptable.

The introduction of procedures SOP, and the REX by making visible a hidden part of the iceberg have greatly contributed to this improvement, but often also to its increase in complexity, making the task of the operator harder and harder to accomplish in the allocated time.

It is sometimes followed by deviations, mostly produced in order to balance safety versus production mentioned earlier.

Thus arises a paradox: increasing visibility makes previously unknown faults appear. To cope with this, procedures become more complex increasing deviations to ensure this balance between safety and production.

It is thus time to invent something new that will leave more space and freedom to the intelligence of the front-line operator, so he can act in an acceptable space, probably supported by a skeleton of essential basic procedures.

If one day we take this way no doubt that the flight simulator will be a special field of experiment.

Appendix

Situation awareness	Good practice	Poor practice
Awareness of aircraft systems	Monitors and reports changes in system states	Does not ask for system updates
	Acknowledges entries and changes to systems	Does not signa awareness of changing systems
		Driven by the system, linked to habits
Awareness of external environment	Collects information about the environment	Does not acknowledge—nor repeat ATC directions
	Contacts external resources when necessary	Does not enquire about environmental changes
	Shares information about the environment with others	Does not comment on relevant environmental factors, or is surprised by them
Awareness of time	Discusses contingency strategies regarding time	Does not set priorities with respect to time limits
	Identifies possible—future problems/forecasts milestones	Does not discuss relationship between past events and present—future
		Is surprised by outcomes of past events

Decision making	Good practice	Poor practice
Problem definition and diagnosis	Gathers information and identifies problem	Nature of the problem not stated or failure to diagnose
	Reviews root causes with other crew members	No discussion of probable causes
Option generation	States alternative courses of action	Does not search for information
	Asks crew members for options	Does not ask crew for alternatives
Risk assessment and option selection	Considers and shares risks of alternative courses of action	Inadequate discussion of limiting factors with crew
	Talks about possible risks for course of action in terms of crew limitations	Failing to inform crew of decision path being taken
	Confirms selected course of action	

References

Amalberti, R. (2001). The paradoxes of almost totally safe transportation systems. *Safety Science*, *37*(2–3), 109–126.

Fauquet-Alekhine, Ph., Labrucherie, M. (2012). Simulation training debriefing as a work activity analysis tool: the case of nuclear reactor pilots and civil aircraft pilots. In Fauquet-Alekhine, Ph. (eds.), *Socio-organizational factors for safe nuclear operation* (1st ed., pp. 79–83). Montagret: Larsen Science

Nixon, P. G. F. (1976). The human function curve. *Practitioner*, *217*(765–769), 935–944.

Reason, J. (2008). *The human contribution: Unsafe acts, accidents and heroic recoveries*. Farnham (UK): Ashgate Publishing Ltd.

Rizzo, A., Bagnara, S., & Visciola, M. (1987). Human error detection process. *International Journal Man-Machine Studies*, *27*, 555–570.

Yerkes, R. M., & Dodson, J. D. (1908). The relation of strength of stimulus to rapidity of habit-formation. *Journal of Comparative Neurology and Psychology*, *18*, 459–482.

Piloting Nuclear Reactors

Philippe Fauquet-Alekhine and Carole Maridonneau

1 The "How" and the "Why" of the Simulation

The operation of nuclear power plants requires a high degree of control of facilities, either in terms of piloting or maintenance, or in normal or accidental situation. It concerns the security and the health of populations, and therefore the possibility of maintaining the nuclear sector in the energy market. The nuclear operator must thus be able to not only maintain its know-how but also to update it in order to adapt to new imperatives, which can intervene at the level of safety or security, regulation or legislation or economy (Buessard and Fauquet 2002; Fauquet-Alekhine 2012a). These imperatives are mainly of two types:

- External: developments in the requirements of the prescription, of nuclear safety that the operator as the nuclear regulator (the safety authority[1]), keeps on strengthening.
- Internal: corrective actions from the analysis and the treatment of events essentially identified by the operator.

---

[1]Autorité de Sûreté Nucléaire (ASN).

Ph. Fauquet-Alekhine (✉)
Laboratory for Research in Science of Energy, Montagret, France
e-mail: philippe.fauquet-alekhine@edf.fr
URL: http://www.hayka-kultura.org

Ph. Fauquet-Alekhine · C. Maridonneau
NPP of Chinon, BP80, F37420 Avoine, France

C. Maridonneau
Montagret, France
e-mail: c.saphore@orange.fr

© Springer International Publishing Switzerland 2016   59
Ph. Fauquet-Alekhine and N. Pehuet (eds.), *Simulation Training:
Fundamentals and Applications*, DOI 10.1007/978-3-319-19914-6_3

Among other means implemented to respond to this ability to maintain the know-how (Fauquet 2003, 2004; Klein et al. 2005; Fauquet-Alekhine 2012b) in a context of skills drain while the inflow of newcomers in the company decreases drastically the average age of experienced workers in teams (Fauquet-Alekhine and Daviet 2015), EDF[2] has developed full scale simulators for training pilots of nuclear reactors[3] (Fauquet-Alekhine 2012c; Fauquet-Alekhine and Daviet 2015) and addresses the sensitive question of newcomers' expectations in terms of learning means and methods and in terms of innovative training methods (Le Bellu and Le Blanc 2012; Fauquet-Alekhine 2013a, b). Control rooms are reproduced in scale 1 (simulators said "full scale"), and calculators allow real-time simulation of the physical parameters of the installation. The choice of such a pedagogical tool is motivated by a dual need: create closest situations possible to real operating situations in order to train a team to pilot a complex socio-technical system collectively, where a complex socio-technical system is defined as a system that combines multiple technical solutions to fulfill a given need (complex technical aspect), the achievement of which requires women and men at work within an organization which is also complex (complex social aspect) (Fauquet-Alekhine 2012a). In this perspective, the full scale simulator has demonstrated its added value for nuclear industry, and years before, for aviation.

The simulator consists of a control room, identical replica of the operating reality (Fig. 1), a calculator, and a control panel from which trainers manage the simulation situation. This desk is closed off by a one-way mirror, equipped with digital video system connected to multiple cameras allowing various views of the control room, with a capacity of zoom such that the reading of a sheet of paper laid on the table of the operators is possible from this console. All of the views of cameras are recordable and can be viewed later in the debriefing room. This video system presents an undeniable added value, as this will be illustrated below. But there are a few drawbacks sometimes difficult to anticipate as long as they are not experienced. Trainer tells: "I realized that when I had zoom-plans on some parts of the control room, sometimes people came between the camera and the object viewed: with auto-focus, the update is done automatically with the bird's-eye view of the person as the cameras are on the ceiling. It highlights then video views of people for themselves with which they are not used and which, sometimes, affect self-esteem: a brilliant baldness, a distort collar, a tag out of pull-over… People, watching the video, pay attention to this kind of detail. We saw a debriefing blocked, following the view of video clips showing the bald head of one of the trainees. Discussion turns short very quickly: people focus on this point and work is put aside. The video is a revelation that can be forthrightly disturbing." The long shots prevent this type of drawbacks and loss of time, and remain quite effective for the analysis of collective practices.

---

[2]Electricité de France.

[3]There are full-scale simulators for maintenance teams and for operating teams; every plant of the nuclear fleet has both.

Fig. 1 Example of a full scale simulator for nuclear reactors

The video must therefore be used with caution, in respect of persons, knowing that this respect for people is not easy to understand with all its subtlety.

It is clear that the only simulator does not reproduce a work situation aiming at being the most realistic possible, i.e. reproducing the better the reality of industrial operating situation. The simulator is only a support tool recreating an environment and interactions between the workers and between the workers and the industrial process that obviously leads to miss some of the interactions (regarding social interaction loss, see Fauquet-Alekhine 2013c). In order to create a simulated situation, scenarios must be ready to engage trainees in action. These scenarios incorporate the input parameters for calculator simulator, and also the input parameters to be given to the team in order to pilot the system; this permits to introduce the history of the work situation that everyone is going to live on the simulator, and to suggest a technical sequence of the process during the simulated situation. The whole is coordinated by the trainers.

The objectives of training according to their types have already been fully described (see for example, Kein et al. 2005). In short, let us notice that different terms and conditions of work exist on simulator based on the objectives sought; for example: initial professionalization, refresher course, accidental procedure, and development of know-how. It is this latter modality which concerns us here, corresponding to the so-called training session in French "*mise en situation*" that we could translate by "putting in situation". Specifically to this type of simulation, objectives are to identify the intellectual approach and resources mobilized by the individual or collective, to analyze the construction of collaborative and adapted actions deployed by them, often around a problem-solving approach. To achieve these objectives, methods of reflective practices analysis are used. We are here with a view to improving professional practices: this protocol allows us to re-examine what is done by the team, what is not done, what are the strengths and weaknesses, with a perspective of possible transformation. This processing may concern the individual or collective, but also the organization of work in real operating situation.

However, any case of "putting in situation" does not necessarily imply a transformation of this type. Sometimes, collective analysis of the work activity emphasizes practices recognized as effective by all the workers (trainees as well as trainers). The added value does not appear therefore in a transformation, but in the fact that workers get conscious of what has happened, thus contributing towards the anchoring of these professional practices.

For this type of simulator training, piloting team should have already elaborated skills implemented in an industrial situation since it is more to adapt or become aware of these practices rather than acquire knowledge in a situation of simulation.

2   The Roles in Simulated Situations

Simulated situations concern operating team members (pilots) and trainers.

First, we must devote a few lines to the description of an operating team for French nuclear reactor. It is generally composed of fifteen persons who support the operating of a pair of reactors and associated equipment. It is a 3×8 shift job.

Insert 1
The nuclear unit
The perimeter of action of the operating team is not restricted to the operation of a nuclear reactor. To produce electricity, a transformer is needed, connected to high voltage lines distributing current to the users. Between the reactor located in the reactor building and the transformer located at the extremity of the turbine hall, lots of equipment is required: auxiliary nuclear materials to operate the reactor and all of the backup circuits in case of accident, a turbine hall with a turbine that powers a generator, as well as all the systems for conditioning and treating water which alternates between liquid and gas with large variations in pressure and temperature. This set, from the nuclear reactor to the transformer, is called "nuclear unit". According to the year it was designed, a nuclear unit is able to produce 900–1450 MWe. The most recent design is the EPR (European Pressurized Reactor) on the site of Flamanville which must produce 1600 MWe.

Operating team mission is to produce electricity from nuclear energy (Fig. 2) while ensuring the safety of installations: to control the nuclear core process, to ensure that the thermal exchanges are of good quality, to get a stability of the alternator, and especially to ensure containment of nuclear facilities. These imply parameter monitoring and adjustment, periodic tests, in order to suggest actions of maintenance when needed. All this is done in part within the control room (one

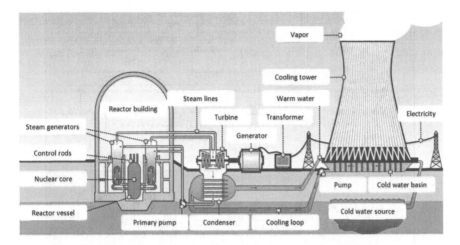

Fig. 2 Simplified diagram of a nuclear unit

room per reactor) as well as locally, directly on the equipment (electric cells, sensors, valves, pumps…).

Insert 2
Interfaces control room-field
These actions are mostly carried out at the request of the operators working in the control room, early on in the shift, by telephone or by direct contact if it is possible for the field worker to be in the control room. Indeed, this is not always the case because, in general, at least two field workers are assigned to the activities of "controlled zone", i.e. in premises potentially dangerous from a radiation protection standpoint. These field workers thus start the shift after going through special dressing rooms and putting on adapted clothing. They come back to these dressing rooms at the end of the shift to join the rest of the team to the shift-team debriefing. In the meantime, any communication with people "outside the controlled zone" is done by phone.

Each team is managed by an operating chief manager (CE on Fig. 3) supported by two deputy managers (organization at 1st Jan. 2015). Operating actions are carried out by pilots who are called "Operators", 4 to 6 persons on the two reactors (a minimum of 2 operators are assigned to each control room to pilot the reactor); the job place of these Operators is therefore the reactor control room. The rest of the team is composed of technicians, also named "field workers" because their tasks concern all of the installation; the field workers manage hydraulic configurations (adjusting the hydraulic circuits by handling valves essentially), measurements of physical parameters from sensors and local reading, detections of anomalies and their diagnostics, and possibly participate in the repairs.

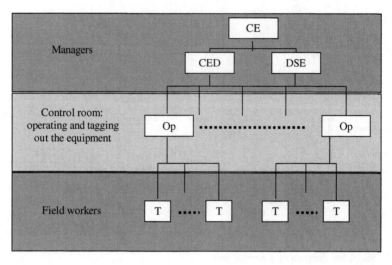

Fig. 3 Chart type of a team (1st Jan. 2015): the manager CE (*Chef d'Exploitation*), with the help of two deputy managers CED (*Chef d'Exploitation Délégué*) and DSE (*Délégué Sécurité en Exploitation*) manage a team of pilots Op (*Opérateur*) among whom two experienced pilots OpPil (*Opérateur Pilotage*, one per control room), and technicians T (*Technicien*)

This succinct presentation of an operating team enables to restrict the population concerned by the piloting simulator: the operating chief manager, the deputy managers, and operators. Indeed, they are the ones who have to act on the equipment of control room, either in terms of decisions, or in terms of actions: if the operators are the only persons allowed to touch the equipment, the team hierarchy can be brought to validate or not the coming actions, or to request the application of a specific piloting strategy.

Simulation situations so called "putting in situation" are therefore a "head of team" dedicated training session.

3 The Situation of Simulation

Going into the situation of simulation (so-called "putting in situation") for an operating team is done on 3 days, each day is broken down into a briefing, a run ("*trace*" in French) on the simulator, and a debriefing session in a room. The briefing lasts less than 30 min. The "run" refers to the actual time on simulator (or simulated situation, with a duration of 2h30–3h). The debriefing session lasts 2h30 (Fig. 4).

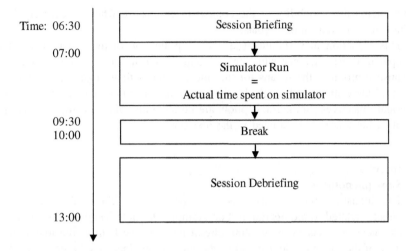

Fig. 4 Example of typical interlocking schedules of a so-called "putting in situation" session

Among the benefits of training on several consecutive days, we can point:

- The possibility to organize a progression on the three days; for example:

  - J1 to analyze the study and resolution of a given problem and to identify the areas for progress,
  - J2 to try to implement what was decided at the previous debriefing.
  - J3 to work the transposition in another situation.

- The existence of a time of integration between two sessions, a time to think (Carpenter et al. 2012).
- The briefing of the session. This briefing time placed just before the session on simulator, helps anchor of new practices discussed during the debriefing of the previous day. This briefing is beneficial for all learning. The production of the previous session remains present in the minds of the trainees and is reactivated by the trainers at this particular time. This re-activates the attention of pilots on the items selected in the debriefing of the day before.

From the point of view of learning and integration of the contribution of the simulation to the forthcoming real operating situation, the simulator in several days together is a real added value.

The situation of simulation is based on the technical tool that is the full scale simulator. As we said, it is clear that the single tool is not enough: it is necessary to build scenarios capable to achieve the objectives, and therefore place trainees in a specific situation about which we shall come back later. These scenarios are elaborated from real-life situations, which we can designate "reference situations". From a reference situation, the scenario permits to develop the simulated situation by reducing the variability of the context to emphasize what is necessary to the achievement of educational goals.

This reduction in variability (i.e. the simplification of the situation simulated compared to the reference situation), is fundamental in the pedagogical approach, because it provides the means for the pilots to focus their cognitive resources on some of the difficulties brought by the script directly in connection with the educational objectives. Sometimes, pilots are complaining about being "too far from reality", but this may be a need in the first stage.

Insert 3
Slow phenomena
The industrial process involves slow physical or chemical phenomena directly related to the operation. For example, the time length of injection of chemical products in the cooling circuit may be due both to the maximum flow of the injection device and to the precautions that may be necessary to ensure a satisfactory diffusion of the chemical inside the whole cooling circuit.

The pressure, temperature, or neutron power are also typically parameters which must sometimes evolve slowly to ensure their control, and therefore the safety of installations.

In real operation, the variation may take several hours. Yet a simulator run lasts only 3 h. It is therefore necessary sometimes, on simulator, to set up a calculation such as the kinetic will be significantly increased in order to work some long-lasting operating phases within these 3 h.

The simulator also allows to vary the kinetic of physical phenomena in two senses: regarding the dilution that would take several hours for example, the trainer is able to speed up the simulation in order to keep the script time limit for a run of 3 h; similarly, for something fast, it is possible to "freeze" the simulator, i.e. freeze the industrial process in a given state so that pilots may take the time to think about what is going on, and the consequences of their actions or their non-actions.

If the benefit of the freezing or accelerating functionality of the simulator (adjust the kinetics of the phenomenon) is immediately visible, we must not forget the possible drawbacks this might induce. Increasing the speed of the physical phenomena does not allow pilots to work the consequence of slow kinetics. For example, the reality of operation induces long waiting periods during which the vigilance can diminish to the point that surveillance of the facility loses in effectiveness: an operator must be able to be trained to this problem which is not done if the simulator is accelerated. By contrast, freezing the simulator does not allow a pilot to work and try to catch up with the immediate consequences of an inappropriate action. Also, varying the kinetics of physical phenomena is not appropriate for the "putting in situation" session. They are adapted to the simulator sessions at technical learning phases.

The approach to the situation of simulation is not only figurative but also operative: to work the impact thereof on the activity of the trainees, the run on the simulator is prolonged by a debriefing in the classroom, which we shall develop further.

In this simulation situation, trainers' place is not neutral. This place is both enveloping and inserted in the situation (Fauquet-Alekhine and Labrucherie 2012).

It is enveloping through their distant observer position which will be essential in the managing of the debriefing.

It is enveloping because they have control of the scenario, by stabilizing or by adjusting the parameters of the simulator. Trainers also provide answers to reactor pilots based on the role they are led to play (only the head of the operating team is trained on piloting simulator: for any hardware simulated intervention, pilots use the telephone to exchange with the maintenance technician, or a field worker for example, role played by the trainer): this is another form of adjustment of parameters of the scenario.

The place of trainers is also inserted precisely because these contributions take place in the history of the temporal interval inside the simulated situation. They may take the role of a field worker of the pilot team, that of a maintenance technician, or voluntarily the disruptive role of any worker in the process.

To capitalize everything that is observed during the run, trainers have several tools which the principal is taking notes.

One of the strengths of the configuration of the simulator is to facilitate taking notes: trainers, separated physically from the pilots, are sitting at a table big enough to receive control computers as notebooks. The physical separation allows an exchange of views in real time between trainers. In addition, computers in control room of the simulator allow recording and printing data showing the evolution of physical parameters that will allow them to illustrate the understanding of a given situation of operating if necessary. Thus, all this history in which trainers participate is meticulously recorded to be used during the debriefing session.

To facilitate taking notes, the trainers have developed computer macro-commands which allow them to launch a series of conventional operations in a few clicks. This frees them from the management of the simulator and allows them to focus on the session and take notes.

Physical separation trainers/operating team presents another advantage: it fosters an involvement of trainees in the situation and a dialogue between trainers without disruption or interference with the members of the team.

However, what trainers may be able to do during the run as well as during the debriefing session is highly dependent on their professional experience. Trainers have various career profiles. Some of them are skilled in the art of piloting the reactors. Being from the profession gives them a legitimacy regarding the persons in training: This provides a facilitated discussion on the technical plan that is felt by the trainees, and allows the trainers to support discussions in debriefing by the narration of experienced examples, understandable for the trainees, and appreciated because it facilitates the understanding of the discussed elements. In addition, this provides a certain attractiveness to the debate, due to the anecdotal character of the story. For new trainers, for people with no operating experience, computer databases containing narratives of events will be investigated before training sessions. This permits open debate also for the experienced trainers who sometimes would tend to stay on their own experience, which, even rich, finds here the benefit of recent operating experience feedback.

One of the counterparts is that trainers coming from operating or maintenance professions can be engaged in technical discussions during the debriefing. The difficulty then rises in the know-how to keep the necessary distance to not fall into this sort of trap. However, the solution cannot be the opposite, i.e. choose trainers who do not have such experience, because this requires access to a legitimacy that is therefore not acquired a priori. There is a need for trainers who are not from the operating profession, to know a minimum which makes the technical basis of these professions. However, they can win legitimacy by the use of techniques or methods appropriate to help trainees to analyze their practices by questioning. The fact that they are not skilled in the art at the time helps to provoke for a productive questioning to the trainees.

The solution for novice trainers without any operating experience may reside in the management of the collective of trainers, attempting to maintain a team of various profiles, by constructing spaces for exchange of practices and working posture during the training of trainers.

In addition to these factors related to the own experience, another point affects the positioning of the trainers: that of their future position based on their previous position.

Trainers are sometimes coming from the operating profession. Remaining in post in the training center for approximately 4 years, they are sometimes asked to return to operating teams on the same nuclear power plant. In the last months of the trainer's job, some of them express their difficulties to speak to colleagues in training: "I can't say it, because they are of former colleagues and I will join them in a few months, a few weeks!"

Again, operating in pairs of trainers helps to deal with the return in the very short term of one of them in an operating team.

A final point remains to be addressed regarding the posture of the trainer. Observations have revealed that some trainers adopt a posture of excessive deference to the trainees: the language they speak, the content of their speech, and their attitude, showed trainers eager to witness a high respect to the functions of the trainees. However this posture is almost systematically coupled with a difficulty for the trainers to conduct the debriefing: they easily accept the counter-argumentations of the operating team, the practices analysis seems to be oriented by the team, and they seem to have difficulty maintaining the discussion on the activity of work as issues are cut short.

This posture is independent of the professional experience.

An opposite posture is also observable (much rarer): coming from the operating profession, trainers expose explicitly their CV and skills, are policy in the debriefing, and submit the solutions instead of working their emergence within the collective analysis, which is counterproductive (Fauquet-Alekhine and Boucherand 2015).

If the first not often can work through the sharing of experience, observations show that the second is hardly discussed.

3.1 Effects Sought and Pedagogical Objectives

The position of trainers as well as their know-how are fundamental in the achievement of the objectives sought when the team is in the situation on the simulator.

From the operational standpoint, on the one hand trainers must be able to manage a large number of parameters from their computers to ensure that their scenario takes place on the basis of the given objectives, and on the other hand, they must be able to manage the direct interactions with the pilots who express solicitations throughout the simulation as they would seek a maintenance technician or a field worker in a non-simulated situation.

In terms of the team work and its development, trainers must be able to adjust these data set permanently to bring and maintain the pilots in a situation of solving problem. It is essentially in this type of situation that trainees will convene their knowledge and develop their skills.

The effect sought in the debriefing session is the distancing of the pilots with their action in the simulated situation that they come to live. To do this, enveloping distant position of an observer which is that of the trainers in a simulated situation is valuable assistance. This distancing with the situation facilitates the understanding of the intellectual approach of the pilots in situation, individually and collectively, and must permit the re-work, i.e. to allow re-thinking to transform. This trainers' distancing also promotes the observation and analysis of the interaction human-process, interactions between people, as well as the individual and collective contributions (or non-contributions) to the action in the situation. The result is a possible transformation of these approaches, interactions, and contributions. We will develop the means and approaches to do this.

All of which is implemented in simulated situations and debriefing is essentially due to the research work of Pastré, Samurçay, and Plénacoste, from 1996 to 2001 (Klein et al. 2005). Béguin and Pastré (2002) describe the conceptual perspective of situations of simulations and debriefings (see also Pastré 2005) which was completed by Fauquet (2007).

To complement this approach, and to assist trainees (as trainers follow their evolution of training from one session to another), a FAP system is implemented. The FAP (in French: "*Fiche d'Aide à la Progression*") is a progress support form drafted jointly by trainees and trainers. They are individual, given to each, and must ensure continuity of the simulation training, regardless of the assessment of the operating teams which is not discussed here. They are reserved for other types of formations such as the refresher course. For the "putting in situation", at the end of 3 days, a contract of collective transference is written. This document is the property of the operating team. Trainers help in its drafting. It contents all the important points observed during the "putting in situation" session: strong team practices (supporting them in their activities on a daily basis) and axes of improvement. Only facts observed and recorded by the team can be included in this document. The team will feel free to use it later. Some teams incorporate the contents of this document in their team project, or use this document as a specific theme work on a day dedicated to the analysis of the functioning of the team.

3.2 The Briefing Session

This 30 min sequence just before the run on the simulator is very important. On the first day, it is the appropriate time to present the basis of the rules for the forthcoming exchanges between participants. It is necessary to establish the limits thereof. The confidentiality of exchanges is necessary. A state of trust must be established, otherwise, participants might find it difficult to give their feelings whereas this contributes positively to analyze their practices.

On the following days, this time preceding the simulator run allows trainers to reactivate the state of alertness of the actors. It is reminding their conclusions from the previous day, their findings. They then undertake implementation of new practices to experiment with them easier. Or even, they undertake to be vigilant about a practice, as for example, always use the 3-way communication (Fauquet 2007; Fauquet-Alekhine 2010; Fauquet-Alekhine et al. 2012) between them during the work, to be discussed in the following debriefing.

This briefing can also reframe the team. The objectives they have on the session are again specified. They must be active to ensure that these sessions are productive for them.

3.3 Going onto the Simulator

The simulated situation always begins with a transfer of information between the trainers and the operating team. This will contain a brief overview of the simulated installation state (current production level, possibly unavailable materials) and the work program provided for each simulated job (change in production to come, planned interventions, periodic testing). The term "brief" is crucial, because it focuses on a first difference with the real operating situation: the 5–10 min thus with trainers are supposed to replace the minimum of 30 min devoted to an exchange with the outgoing team (the shift) and a incoming team bringing together about fifteen persons concerned through a team briefing. From the outset, this first step makes trainees not forget they are on simulator and not in a real operating situation, because the "brief" concerns the time spent for the shift-team briefing, but also the number of people interacting with the trainees, the number of processing problems and workload associated while taking position: for each of these items, the intensity is much lower than in simulated situation than in a real operating situation.

Very often, the trainees explain that they are "here to manage a defect which is to come." They therefore start by watching out the control room looking for the slightest deviation on any indicator in order to detect the earliest this hypothetical defect. Thus, we can sometimes observe trainees in simulated situation focusing on what that seems to be such an indicator, for example, an indicator of level of tank slightly more than normal. Operators will give particular importance in order to see if the value changes no more, proof of abnormal evolution of an operating physical parameter. Thus, focused on this parameter, they give insufficient importance to the monitoring of the installation in a global manner, and take the momentary risk not to detect the effective indicator announcing "the expected defect", or they detect it too late. This has been watch on simulator. Furthermore, in some observed cases, operators can lead the rest of the team in this "tunnel effect".

The role of the trainers can then be to use this context of "tunneling" to lead the team to learn from this in order to apply the lesson to the real operating situation later: indeed, trainers have the choice of the moment where the actual defect indicator will appear in the control room. A possibility for them is to generate this defect at the time where the trainees are the busiest to reflect on "the false defect". The objective is not to give them a good lesson in the type: "you are not there to wait for the blackout but to do as you would do in a non-simulated situation". The objective is to use this simulated situation to make them think the analogy with the non-simulated situation during the following debriefing, by guiding them to see what has been experienced, and showing them the risk incurred by a concentration of team resources on a single problem at the expense of the whole process. In this way, what could appear as a loss due to the attitude of the trainees is transformed into an added value for the desired effects.

In the first 30 min of the run, trainers will induce a disturbance that will invite pilots to identify the indicator and the problem, then to elaborate a diagnosis. Throughout this 3 h simulator run, trainers will try to observe the team piloting the reactor according to 4 points:

- communication,
- application of the procedures,
- collective organization,
- analysis and problem solving.

Trainers live the scenario according to the pace of the operating team in the control room: the more activities the scenario involves the team in, the more their contribution to the management of the scenario is important; the software settings are more numerous, the number of telephone interactions increases. They must manage the verbal exchanges between the team members and others by playing different roles on the phone, sometimes going inside the control room; they must permanently adapt operating parameters according to the actions requested by the team to others; they must ensure the scenario running. Experience shows that this is a problem especially for trainers involved recently in the job: in a first period, they have to integrate the technical operating of the simulator, which mobilizes a large part of their cognitive resources (just as the young driver who focuses to think what he must do in the car before driving it, and who will be able after 1 year to run these commands without thinking, doing something else in parallel).

Thus, all young trainers will go through a difficult period before acquiring the spontaneous response thus decreasing their cognitive load. It is one of the reasons for which it is important that trainers work in pairs, in particular during the "putting in situation".

In such a context, their capacities to observe the team decrease whereas it is precisely during such periods that it becomes relevant.[4] This difficulty has recently mitigated by developing computer macro-commands which allow in a few clicks to launch a series of conventional operations. Conversely, when the atmosphere by the side of the control room is quiet, the trainers have time to observe (both can do it, or one observes and the other manages the simulator and the interactions with the team) but at this time, there is much less to notice.

The co-management in pair is a resource, combined with the fact that trainers are separated physically from the operating team: trainers can discuss freely what has happened on the heavy work periods and integrate it in their notes.

This privileged position of invisible observer allows trainers a review of the postures of each of the trainees in the simulated situation. Thus, they are able to integrate objective elements to their notes to help operating team to better understand what is played out in this story. Trainers will thus create a self-representation of the work activity and will confront it to the one of the operating team during the debriefing session. This confrontation is done by questioning, trying to stay on the "how" and not the "why": "How did you do for...." The objective of such questions is to access both to how pilots have made as to what they could not do. Similarly, the trainers are well placed to identify and analyze the elaboration of delicate situations, such as tensions that arise in the team and may lead to the back off of a person, and to explore the causes. During a "putting in situation" session, the cases of back off do not happen. Nevertheless, this has already been observed during another kind of simulation session and we can assume that one day it will happen during "putting in situation".

Example of observation of a case of back off:
It is the second day of simulation training. The beginning of the session starts with the distribution of roles because sometimes, the trainees want to take advantage of the various sessions over several days to take a different function, for example, to discover another point of view of the work situation. But one of the trainees refuses. The discussion is committed between different people and finally, the roles change except for this person. Trainers observe then, through this person's attitude, that he voluntary puts himself apart of the team. Soon, he is expected to participate in a diagnostic operation which requires several points of view. But he has always something else to do: watching the monitoring of

---

[4]Trainers of the Training Center of Chinon asked in 2005 an analysis of their work activity with the objective to better understand which were their strengths and weaknesses during the debriefing (Fauquet 2005). The analysis found out links between these run phases and the debriefing.

the evolution of a parameter, then writing screen data, with a displacement from one computer screen to another. It looks more like a leak as soon as someone approaches him rather than a real need of information intake from the process. The others finally adopt a conspicuous waiting attitude and ask him to join them. The person joined the group around the desk located in the middle of the control room. However, trainers note how his commitment remains distant, and how the other members of the team seem embarrassed by the tense climate created.

The observation of the simulated situation by trainers help them analyzing how the team is implementing the professional practices, apply the operating modes, knows and understands the operation of the installation. In this, it can be seen that the team is able to achieve in the work activity with regards of what is expected; this is the technical aspect of the work. But it also allows them to point out both right and wrong particularities of individual or collective functioning.

Example of a case of evident reliable practice:
The operator faces the control panel and is about to make a regulation of the scram position. He must adjust the reactor neutron power to the level requested by the electrical network. Before handling joysticks, he says out loud what he intends to do in the presence of the technical manager who is right behind him. The trainer benchmarks this approach as a crossed control; according to him, this professional practice allows to make the action reliable before acting as it is to be seen before performing the action by the technical manager who has a global vision of the state of the installation more wide than that of the operator. Indeed, operators always share binomial control room activity, which contributes on time to limit this knowledge of the general state of the process. However, the trainer does not know whether the operator is aware to implementing a reliable practice or not. He notices therefore this point, identifies the sequence on the video recording in order to have a possibly based re-discussion during the debriefing.

The video recording presents other benefits (Fauquet-Alekhine and Maridonneau 2012) which will complement the collective word of the debriefing session: recordings of passages on simulator to show some actions carried out by the team whose actors are not necessarily aware. For example, it is not uncommon that the team implements reliable practices that seem so natural that they are unable to carry them on spontaneously in the debate; how to spread easily and rapidly such good practices to the novice if trainees are not aware of them. This clearly contributes to anchor these practices in learning. Conversely, the team may have put in place practices that seem reliable for themselves, but for which practices the experience feedback...) shows that they are a reference to the team only: in case of just a replacement in the team, it does not work anymore (cases of implicit terms while exchanging information for example). Video recording permits to demonstrate such strengths or weaknesses (Fauquet-Alekhine 2009, 2010).

Since a few years, the simulator of the Chinon plant is equipped with a digital video system. The added value in terms of exploitation is undeniable, with the possibility of recording multi-camera, and tracking by computer specific sequences, easy to find thereafter.

Example of a case highlighting a management problem in the team:
This is the first day of the team on simulator. The scenario led the team to reduce, for safety concerns, the power of the installation following a technical problem on hardware. While

the diagnosis is done, and while the automatism technicians have been called upon to work on the defect since several minutes, the technical manager takes the phone to call the manager of the electrical network. The operators are then in action in the control room. When the interlocutor takes the phone,[5] the technical manager explains that the installation has a problem which requires reducing power provided to the network from 100 to 86 % of full power. The interlocutor explains the technical manager that he just had the information from an operator of the team and that he is already aware of this. The technical manager hangs up the phone, walks around the control room during the operators go about their tasks, and again seized the telephone to call the automatism team. He asks his interlocutor how long it will take setting of failing components, knowing that it is a required condition to resume the power at 100 %. The interlocutor gives the answer, and explains that he had this conversation a few minutes earlier with an operator. The technical manager apologizes, says that he was not aware and hangs up the phone. Then he goes on his walk around without any remarks to the operators.

For their part, trainers note these two events as illustrative of a management charac-teristic of the team: other elements have shown before that the operators and the technical manager did not share the same representation of the installation state, and especially do not share at the very beginning any agreement between them regarding the strategy of treatment of the difficulties. This was made visible by the induced problems. The trainers make the assumption that the technical manager does not or cannot perform his task of coordination and they plan to bring the discussion on this point during the debriefing.

To catch such events and use them in debriefing, trainers need to share the understanding of what they observe because they each have their own represen-tation of what happens in the control room. This is why they permanently create exchange points between them during the trace. The observation of their work activity (Fauquet 2005a) permitted to characterize these points: they discuss what is going on (regarding the technical level of the operating simulation) and how they perceive the situation experienced by the team. A short exchange between them every 3–5 min is noticed. This allows them to share their representation of the simulated situation. However, more than 10 % of these exchanges are interrupted by requests (including interactions with the control room), which implies a mental effort to resume the exchange, hence a loss of efficiency or, rarely, of the previous conversation.

This type of exchanges permits to strengthen the identification of features, such as trivial gestures that might not be so common. This is the case of a single sheet of paper on the desk.

Example of the sheet of paper:
One of the operators is underway of a periodic test on a hydraulic loop while his colleague manages the monitoring of the installation. He must act from a desk on the command equipment. One of the trainers notes that the operator takes a sheet of white paper, in addition to his procedure, and puts it on the console as if he needed to take notes; however, put this way on the console, it is impossible to write correctly on the paper. This feature

---

[5]The interlocutor is a trainer: s/he answers the phone from the simulator control panel and plays a role according to the situation, trying to integrate as many elements as possible recalling the real operating situation in her/his speech, in what s/he suggests or refuses. For example, s/he may explain being busy and performing another activity; s/he may explain s/he will do what is asked but that it will take time because s/he is so far from the place where must be done the work.

calls out one of the trainers who communicates it to his colleague: no explanation; they note this point for further discussion during the debriefing. This is a practice (conscious or not) that it will be interesting to put into discussion at the debriefing session.

Among the four axes of observation of the trainers, one addresses communication. It is ubiquitous in the operators' activity because they are almost permanently interacting with others: they pilot in pair the installation from the control room, request support of their field workers to carry out actions in connection with their activity in progress, they are interacting with maintenance workers to manage diagnostics of power outages or repairs, and for the re-qualifications. During these verbal exchanges, the operators are engaged to make requests for specific equipment of the installation; for example, a valve has to be opened, a pump has to be started, local parameters must be checked (physical values such as temperature, pressure or tank level). To allow a full identification of the equipment, a nomenclature of functional marks has been established at the design stage of the facilities; it is made as follows:

reactor number + system + equipment number + type of equipment

Thus, 1 RRI 002 PO refers to the pump (by "PO") number 2 (by "002") of the RRI system on the reactor number 1. However, these functional marks are perceived as "long", and repeating it several times in a sentence is felt to be tedious; it seems even that saying once in the sentence is not obvious to the issuer of the message. Also we can hear "pour R-R-I, tu me démarre la 2 P-O" (for R-R-I, you starts the two P-O) where one might expect: "tranche 1, tu démarres maintenant la 1 R-R-I 2 P-O" (reactor number 1, you start now the 1 R-R-I 2 P-O). Indeed, unlike civil aviation, since few time there is verbal communication protocol which is gradually included in professional practices for operations presenting particular stakes. Until 2010, just the management of accidental situations has always had structured communication between workers by the use of procedures which are concerned by specific trainings.

The analysis of operating events impacting nuclear safety suggests that a better formulation of requests for action of this kind would reduce events occurrence. Simulation situation, regardless of the scenario chosen, helps trainers working this transverse aspect of the work activity: on one hand, a protocol was established (to express the request for action by mentioning functional marks, action, time, and to require a repetition that must be confirmed), on the other hand, trainers take situations of verbal exchange to highlight them in debriefing if they are reliable, or to work on them in real time if necessary. Trainers thus implement different strategies.

Examples of a trainer's strategy to work on requests with reliable (or 3-way) communication:

- The trainer suggests the use of communication protocol by applying it during the telephone exchanges with trainees; for example, s/he repeats the request that the operator makes for her/him, or makes her/his interlocutor repeat. It is clear that trainees build their following exchanges on this model.

- S/He plays the role of the field worker misleading work place (the wrong reactor number for example) if this information has not been specified in the request, which urges pilots to take cautions when following requests.
- When the communication is indeed reliable, trainers identify the video sequence. The use of this recording during the debriefing will allow participants to view the exchange to ensure that the trainees were aware of what they have done. Indeed, it is not certain that this implementation is conscious; in addition, the discussion in the room allows participants to make the demonstration of the benefit obtained in the work activity for the team.
- Similarly, the recent communication protocol specifies that any application for action must be repeated and confirmed:

  - *Tu ouvres la vanne* 1RIS75VP ! (you open the valve 1RIS75VP)
  - *J'ouvre la vanne* 1RIS75VP. (I open the valve 1RIS75VP)
  - *Ok, c'est correct.* (Ok, this is correct)

The video will allow pilots to see during the debriefing session whether or not the communication is consistent with this protocol.

3.4 The Debriefing Session

A break of half an hour is intercalated between the run and the debriefing session. Do not neglect the pedagogical function of this break, beyond its character of recovery. Indeed, while trainers prepare the debriefing, this time-break allows the operating team to relax and discuss a number of problems. In general, it comes to technical problems, which is beneficial for the debriefing that will follow: this aspect being partly discussed in advance, it is easier for the trainer to focus the discussion on other issues by avoiding a technical focus.

Then follows the debriefing. Its length is 2h30 to 3h: it allows substantive work on the analysis of the practices and the simulated situation. Experience shows that this duration is well suited to allow collective analysis of the run that lasts 2h30 to 3h. In addition, in case of particular difficulties, such duration allows to have time to treat a confrontational point and then make all trainees available for the debriefing continuation. Also we should keep in mind that a simulated situation has meaning only by combining durations run/adapted debriefing, where "adapted" implies that time debriefing should give time for discussion to what is done during the run. In other words, believing that a good simulation situation is one that favors the simulator time is an error.

The trainee/situation relationship aims at producing knowledge which is subordinated to an intra and interpersonal psychic dimension: the operator has to examine what has been produced or not, from the individual and the collective standpoints (Fauquet 2005b). At this stage of the simulated situation, the role of the trainer is fundamental as a help to the query expression and production of answers through an account of what has been lived: this is the "retrodiction".

To initiate the retrodiction of the trainees during the debriefing, one of the techniques often used by trainers is to start by asking feelings regarding the run on simulator, and then moving to the reporting of what has happened.

Trainers use media tools to promote the expression of the pilots' feelings while reducing the bias induced by the influence of colleagues; for example:

- They require the operating team members to give a mark from 0 to 10 on a paper, individually, as an assessment of their perception of the functioning of the team on four axes: team organization, analysis and problem solving, communication, application of the procedures.
- They require to characterize each of the four axes by a representative "weather" of the meeting.
- They require to characterize each of the four axes by a single adjective, difficult exercise for which people give several adjectives instead of one per axis.

The collection of results then permits the discussion on the basis of a comparison of each one's representations.

This also ensures the expression of all members of the team regardless of the hierarchical level. Because often, the risk is that the hierarchy should take the floor first, and that subordinates do that grant or rein their own expression. To avoid this, the floor is first given to subordinates by asking others to intervene. Similarly, if the trainers to identify a person whose charisma or leadership on the embarrassment the expression of the other team, it will be invited to speak last.

Starting with the expression of the trainees' feelings has an advantage, that to discuss possible interpersonal difficulties observed in a simulated situation. Because if that is the case, the trainees will not be available to discuss something else: they must make their mind free of it. This type of situation occurs rarely; however, experience has shown that this kind of difficulty requires at least 15–20 min of collective discussion and analysis of the conflict situation for the trainees to recover their availability to discuss something else.

Example of a case of back off:
The trainers have observed how one of the workers has voluntarily put himself apart of the team at the beginning of the session on simulator. By his attitude, he seems to have impeded the functioning of the collective. We have seen above the description of this case during the run on simulator. Let's see what it gives in debriefing.

After a break of half an hour, the debriefing begins with the expression of trainees' feelings. The first two workers express their facilities or difficulties to understand the physical phenomena, to manage the operation of the equipment. Then comes the turn of the individual who backed off from the session: "I have nothing to say," he calmly says, "it is ok." Trainers attempt to obtain more, in vain. Then the last two people speak out on the same points than the first; no comment on the interactions between people, on the collective organization of the work, while it had been discussed during the debriefing of the first day, as if it is difficult to put this point in discussion.

Trainers ask then to tell about what happened early in the session. The more loquacious explain, and from time to time but systematically trainers require the back off worker to speak about what he was doing at that time there. Then comes for this trainee a question from a trainer: "and then, in your opinion, what does this produce on the work of your colleagues. The person does not know but suggests asking colleagues. With precautions, the other team members explain the discomfort felt due to the attitude of the back off worker. If necessary, the trainers objective facts by relating their observation, and the person seems to be aware of what his attitude has generated. The parallel was made by the team, at the request of the trainers, with a real operating situation and the potential consequences on the correct running of operations. The person who backed off gives his own analysis of the situation, finally agrees that it is not so good, or even potentially dangerous, and accepts to behave otherwise now. From that time (about 30 min after the start of the debriefing), the person is no longer indented in the discussion and his attitude changes drastically. The work of collective analysis of the work activity then takes another magnitude.

Expression of feelings may also be engaged by issues such as: "ok then, I propose each one to tell us how you have been living this session, what is your feeling, how did you feel yourself in the collective, and what did you think of this collective."

While everyone expresses what oneself thinks, the discussion shifts to the chronology of what happened. Each trainee then makes a narrative contribution by small keys to transform the first disorganized narrative in a story that makes sense. But makes sense at the surface only, i.e. that it returns to what is immediately visible. In this "arrangement on the surface of the story", it is useful to know how to get the details of the work activity, without to funnel in the field of pure operating technique, as would do a guidebook.

The role of the trainer, as a catalyst for the speech, then becomes fundamental in the development of the narrative: s/he must lead the trainees to explain what they did, to rephrase, to understand how each of them approaches the situation beyond what is a priori agreed. S/He gives access to another part of the story, still hidden by this first "arrangement on the surface".

Let us consider again the case of evidence of a reliable practice and the case highlighting a management problem in the team.

Example of evident reliable practice:
When the operating team is ask to detail what was done for the neutron power adjustment, the operator who was piloting control rod assemblies explains his technical gesture, what he did so by laying his hands on the joysticks. The trainer encouraged him to tell what was just before; the operator explains that he analyzed the situation to choose what action was adapted, he announced to his colleague operator the planned load decrease, and that he manipulated the control equipment. Is this all? The operator thinks "Yes", it was the main things. The trainer then asked the other trainees where they were at this time. The technical manager remembers having been just behind the operator. The trainer is therefore asks again his question: what happened before the action on the joysticks? The technical manager remembers how the operator explained to him what he was going to do, and the details return to the operator's memory. The trainer suggests a deeper exchange between the participants by the questioning of these elements of activity apparently so common that they became invisible. This is for everyone to put into words what has been done, which was brought by each for oneself and for the other, and what each withdrew from what did

the other. The conclusion of the team is clear: for the power setting, when the operator said out loud what he intends to do in the presence of the technical manager just behind him, this contributes to make what will go on after more reliable: discussing between workers upcoming actions anticipates the risk of error. The trainer then leads them to wonder about the difficulty they had to remember these "details": the general answer is that "it is natural" and therefore nobody thinks about that anymore. The trainer is continuing the collective discussion of the recurring character of this type of sharing prior to the action.

In this type of debate, the trainer brings trainees to re-examine what is agreed in practice, and encourages pilots to set their personal style. It allows them to (re)make conscious what they are implementing in the work activity, and possibly to make it available for others: being conscious on order to transmit to others.

The trainer can also induce pilots to speak about the usefulness for oneself to adopt such a practice. This leads them to become conscious of their practices, and possibly to be aware of them for other situations. Being conscious, individual first, broadcasts practices in the collective to fit into a professional genre.

Example of a case highlighting a management problem in the team:
In this example, the technical manager has twice carried out an action that had been made by an operator just before him. The trainer asked the team to tell the details of the situation and to explain why, according to them, the technical manager could be in this situation. The team re-questions its global operation, both before and after the event. According to them, the technical manager has a coordinating role in the collective work ("he was a pivotal member of the team, it is unavoidable") and this role was played by one of the operators in a manner more asserted during the run. The trainees proposed that the technical manager has failed to take the information necessary for the coordination of the functioning of the team (hence inducing redundant actions); unconsciously, the collective has offset this dysfunction through the implementation of working arrangements permitting nevertheless to continue the work activity. The trainer highlights the gap between the functional schema expected (where the technical manager is the pivotal member of the team) and actual communicative scheme (where the pivotal member is one of the operators): this gap does not make the best use of all resources of the team, and everyone agrees. He then asked the team to propose a functioning solution for the next day which could put these two patterns in consistency. Participants expressed the need of "grouping points", consolidation time-breaks in the action where all gather around the desk in the centre of the control room to share their representations of the situation and the strategy of future operating. Consolidation time-breaks are short: 1–2 min.

This example illustrates how, during the debriefing, the work of collective discussion is an actual possibility of questioning the pilots' profession in the generic sense, using the benefits of each. As explained in the last chapter, it is the personal style that questions the professional genre. It also illustrates the importance for the trainees to produce the solution regarding identified difficulties by themselves rather than being told by the trainers. This so-called "generation effect" is crucial in the learning process (Fauquet-Alekhine and Boucherand 2015) as generative and adaptative (Proctor and Gubler 2008).

Personal style is also reflected by a priori innocuous acts. This is the case of the sheet of white paper put on the panel during a periodic test.

Example of the sheet of paper:
One of the operators, in progress of a periodic test on a hydraulic circuit, has put a white
sheet of paper which cannot be used to take any notes. Indeed, no note will be laid on it.
Trainers return this observation in debriefing as part of their astonishment. The operator
explains that, as with many other circuits, one on which he must intervene is duplicated:
therefore control equipment of the line A and line B exists on the control panel for this
circuit. His actions must focus only on line A. To be sure not to err, the operator places a
sheet on the equipment of the line B. Doing so, he reinforces the reliability of his actions.

   The discussion of this practice gives to think about for the other operator, who did not
do so but agrees that this may give benefits. Then a discussion comes on these innocuous
practices, consisting of diverting an object from its daily use to do something else for the
work activity. Two other examples are then cited: a control room operator was seen placing
a small piece of paper salient on the board of the REA system control to remind himself that
a backup is in progress and that his attention is required (by a circular look in the control
room he will be attracted by the paper and, therefore, even if he was distracted by other
requests, it reduces the probability of forgetting that a backup is in progress); another
operator was seen leaving open the hood of a recorder while they are all closed (this allows
him to remember that there is a specific physical parameter to follow on this recorder;
similarly, a circular control room look will put him back to monitoring this parameter).

The discussion of these "tricks" of the trade participates in their dissemination
via the debriefing of the simulated situation, and contributes to revisit the professional genre.

The way in which these "tricks" are described is fundamental: the worker do not
explain by why he does this but how it does. Explaining the "how" helps to transmit
knowledge and to suggest know-how and expose it to the debate; explaining the
"why" would likely explain mostly the prescription. Instead, what allows the work of
the work, of the professional genre, is such collective debate about the skills of each.

3.4.1 The "Simulator Effect" Invoked

Which trainers have not been confronted at some stage of their career by the classic
answer "it is the simulator effect"? Trainees sometimes invoke this type of argument to explain their actions on simulator for which observation highlights a
deviated act or attitude from what is expected. The most common case is the
attempt to explain the exchange of information giving rise to inappropriate actions
in the situation: the trainees explain for example that the field worker (role played
on the telephone by a trainer) made a mistake in the field not because the application was incomplete or poorly worded, but "because it is the simulator" and hence
the trainer did not incorporate the request as a person of the team would have done:
"my colleague, the real one, he is used to doing it, he would have understood
immediately; no need to explain for a long time". According to the trainees, if it had
been the "true field worker" on the phone, it would not have happened so because
their field workers know what has to be done even with an incomplete message.
They thus refer to an agreed implicit in their team.

   Yet, work analyses have demonstrated that this argument is not admissible.
Indeed, the observations made in control room, both simulated and non-simulated

situations, combined with event analysis, showed that implicit communication are develop in any operating team (Fauquet 2006; Fauquet-Alekhine 2009, 2010). However, if there is a kind of common core from one team to another, subtleties exist from one team to the other that make this "standard of communication" specific to a team. In addition, in a given team, this standard may vary from one colleague to another; for example, implicit used in the control room may differ from those used in the field. This is how some event occurs, disturbing the safety of the installation.

Example of event impacting the safety of the installation due to specific standard of communication:

The control room operator wishes to re-qualified the end stroke on a valve of the RIS system. He calls a field worker and asks him to come in the control room in order to clarify the request. This field worker has just been working with him about a periodic test of the EAS system.

When the field worker arrives in the control room, the operator tells him that there is a problem on the RIS system. Then he leads him to the control panel of the RIS system, which is adjacent to the panel of the EAS system. The operator explains what he wants to do, why he wants to do it, and what is the role expected of the field worker. In this explanatory phase, the "RIS" term is not any more pronounced; only the equipment number and type are used (for example, to designate the pump no. 1 of the system RIS, the operator says "the one P–O"). Why? Because according to the standard of people used to working in the control room, it has long been to do as follows:

1. to mention only once at the beginning of the exchange the system on which relates a conversation,
2. to move to the panel related to the system about which is the conversation.

However the analyses have shown that it is not the case for all field workers, and in this case, is not the case for the field worker involved in this event. Thus, when the operator asks him to turn off "the one P–O", the field worker is leaving the control room thinking about EAS system. Among other elements to understanding this confusion, besides the communication aspect, the field worker has made amalgam with the actions which had been requested previously on the EAS system. He leaves and put therefore off-voltage the pump no. 1 of the EAS instead of RIS system.

This event, example among others, illustrates the risk of basing the exchange of information on implicit terms, knowing that they may concern the teams by parts. In this context, what happens when people assigned to a team are replaced by people of another team?

What highlight the operating teams as a benefit for the work, a way to be effective, only works for a given group of persons, and becomes a weaknesses as soon as this group incorporates an "intruder". Or teams must be able to work reliably even with replacement members.

Similarly, simulator, members of the operating team must be able to be understood by the trainers acting as a field worker. Thus, to argue "simulator effect" such as presented above to explain the occurrence of an inappropriate action is not admissible. Trainers can even "funnel" into the breach offered by trainees on the simulator as soon as the opportunity arises, specifically to elaborate the pretext of a debriefing discussion addressing the risks of implicit communication.

3.4.2 The "Simulator Effect" Operated

Another explicit manifestation of "simulator effect" is the general attitude of operating team at the beginning of the run. As already indicated at the beginning of the chapter, trainees often explain during the debriefing that at the beginning of the session, they scour the control room more to seek the expected defect to fulfill their role of monitoring and piloting: "we know that we will have a problem since we are on the simulator and there must be a defect to make us think; then we expect it, we look for it. Earlier it is found, and better it is managed."

This attitude sometimes leads to situations that do not allow the team to identify the so long-awaited defect. Indeed, sometimes the team wishes so much to identify a defect that they invent it. For example, an operator believes he detects a too low value on the probe indicator related to a tank level, talks about that to his colleague who validates the need to investigate this issue. The two operators share their concern with the technical manager who also involves himself in the reflection, probably motivated by the fact that the alleged defect is perceived as a problem by two people. When information comes to the operating chief manager, there are already three people who have identified the same problem, what suggests him that the problem is real. In short, all are focused on the "false problem" when the real problem happens, which is detected late. It is the "tunnel effect" referred to the Sect. 3.3.

This type of behavior of the team can be observed on simulator, later in the scenario, at a time when all of them are caught up in the story, and forget the simulator. Their desire to solve the current problem may lead them to not detect the next defect quickly.

Thus, wanting to be more efficient in the treatment of a defect, the entire team sailed in a strategy that leads to the opposite of the intended objective.

When this occurs, the trainers know how to take advantage of such situations to help the team to get the most out of it and thus use it on a daily basis. Indeed, these situations are well suited to illustrate the risk linked to the focus of a team on one point at the expense of the overall supervision of the installation.

The briefing prior to the run must allow the trainer to operate this type of difficulty relative to the "tunnel effect": the second day of "putting into situation", trainers will ask the team to conduct what they intend to implement in connection with this point observed in J1, by entering it in a continuity of the previous session debriefing. Trainers activate the pilots' special attention on some of their practices. The analysis that they perform during the subsequent debriefing will teach them even more. This approach does not rely on the technique of the industrial process but on the awareness of the demonstration and analysis of practices, focusing on the "how" more than on the "why".

4 Concluding Remarks

Proposing a team to perform work in simulated situation contributes towards making work activities more reliable as practices are re-questioned, re-thought for new individual and collective development. This point is fundamental to the management of industrial high risk systems while research shows that this management tends to migrate to operation areas less secure than provided by the designer originally. This type of migration, well described by De la Garza and Fadier (2007), can be induced, among others, by ignorance of certain risks, operation and production constraints, and tolerance of the organization to accept exceeding certain limits. "Putting in situation" on simulator allows to re-examine the appropriateness of the terms of actions implemented by the workers in such a socio-technical system.

It is a dynamic process, but slow. Operating team cannot advance on its practices only if workers have the will to do so. To get them there, "putting in situation" on simulator is an effective tool, with all its technical and human factors subtleties: sometimes, it is difficult for the working group, in debriefing session, not to funnel into a technical debate where each excels.

This adds another difficulty: one related to the fact that speech unveils one individual to another in an intimate manner; however any embarrassment of a person can block collective production. On this point, trainers must have good receptivity.

References

Béguin, P. & Pastré, P. (2002). Working, learning and design through simulation. *Proceedings of the XIe European Conference on Cognitive Ergonomics: Cognition, Culture and Design.* Catalina, Italy. September 5–13 2002.

Buessard, M.-J., & Fauquet, Ph. (2002). Impact de la prescription sur les activités de travail en centrale nucléaire. *Proceedings of the 37th SELF Congress, Aix-en-Provence, France* (pp. 326–335).

Carpenter, S. K., Cepeda, N. J., Rohrer, D., Kang, S. H., & Pashler, H. (2012). Using spacing to enhance diverse forms of learning: Review of recent research and implications for instruction. *Educational Psychology Review, 24*(3), 369–378.

De la Garza, C. & Fadier, E. (2007) Le retour d'expérience en tant que cadre théorique pour l'analyse de l'activité et de la conception sûre. *activités, 4*(1), 188–197, http://www.activites. org/v4n1/v4n1.pdf.

Fauquet, Ph. (2003). Analyse de risques des activités de travail en centrale nucléaire : du contexte de l'apprentissage à l'application. *Proceedings of the 38th SELF Congress, Paris, France.* 636–646.

Fauquet, Ph. (2004). Importance of decentralized organization for safety sharing. *Proceedings of the 11th Int. Symp. Loss Prevention & Safety Promotion in Process Industries,* Praha, CZ. 1378–1380.

Fauquet, Ph. (2005a). *Synthèse d'observations et d'analyses de l'activité de travail des Instructeurs Simulateurs. Note d'étude du Centre Nucléaire de Production d'Electricité de Chinon,* EDF Report reference: D.5170/DIR/NED/05.003.

Fauquet, Ph. (2005b). Applied crossed confrontation for context evolution. *Proceedings of the 5th International and Interdisciplinary Conference on Modeling and Using Context*, (pp. 36–41). Paris, France.

Fauquet, Ph. (2006). *Phase expérimentale relative à la Communication Opérationnelle Sécurisée – Résultats 2005*. EDF Report reference: D.5170/DIR/NED/06.001.

Fauquet, Ph. (2007). *Développement des pratiques de fiabilisation sur simulateur de pilotage de réacteur nucléaire* (pp. 129–135). Risques industriels majeurs, Toulouse, France: Actes du Colloque de l'Ass. Int. des Sociologues de Langue Française.

Fauquet-Alekhine, Ph. (2009). Надежность рабочего сообщения для операторов ядерных. реакторов: изучение на тренажерах, анализ случаев и укрепление безопасности. *XXXIIe Coll. Int. de Linguistique Fonctionnelle*, Minsk, pp. 207–210.

Fauquet-Alekhine, Ph. (2010). Use of simulator training for the study of operational communication—the case of pilots of French nuclear reactors: reinforcement of reliability. *Annual Simulation Symposium (ANSS 2010), 43*, 216–221.

Fauquet-Alekhine, Ph. (2012a). Safety and reliability for nuclear production. In Ph Fauquet-Alekhine (Ed.), *Socio-organizational factors for safe nuclear operation* (1st ed., pp. 25–30). Montagret: Larsen Science.

Fauquet-Alekhine, Ph. (2012b). Retraining for manager in the frame of HP program: The case of Chinon NPP. In Ph. Fauquet-Alekhine (Ed.), *Socio-organizational factors for safe nuclear operation* (1st ed., pp. 58–62). Montagret: Larsen Science.

Fauquet-Alekhine, Ph. (2012c). Simulation for training pilots of French nuclear power plants. In Ph. Fauquet-Alekhine (Ed.), *Socio-organizational factors for safe nuclear operation* (1st ed., pp. 69–74). Montagret: Larsen Science.

Fauquet-Alekhine, Ph. (2013a). Information an communication technologies vs education and training: Contribution to understand the Millennials' generational effect. *Proceedings of the International Conference on Electrical, Computer, Electronics and Communication Engineering* (pp. 358–363). Venezia, IT, 80.

Fauquet-Alekhine, Ph. (2013b). Learners' behavior in teaching context: Characterizing the use of information and communication technologies. *Proceedings of the 1st International Science and Applied Research Conference Social Psychology in Education Space* (262–264) Moscow, Russia.

Fauquet-Alekhine, Ph. (2013c). Virtual training, human-computer and software interactions, and social-based embodiness. *Proceedings of the International Conference on Electrical, Computer, Electronics and Communication Engineering* (pp. 364–370) Venezia, IT, 80.

Fauquet-Alekhine, Ph., & Boucherand, A. (2015). Optimal protocol for debriefing of simulation training session (submitted to *Simulation & Gaming*).

Fauquet-Alekhine, Ph., & Daviet, Fr. (2015). Detection and characterization of tacit occupational knowledge through speech and behavior analysis. *International Journal of Innovation, Management and Technology, 6*(1), 21–25.

Fauquet-Alekhine, Ph., Boucherand, A.., & Lahondere, Y. (2012). Anticipating and reducing risks on nuclear industrial plants. In Ph. Fauquet-Alekhine (Ed.), *Socio-organizational factors for safe nuclear operation* (1st ed., pp. 53–57). Montagret: Larsen Science.

Fauquet-Alekhine, Ph., & Maridonneau, C. (2012). Using audio-video recording on simulator training sessions: Advantages, drawbacks, and dangers. In Ph. Fauquet-Alekhine (Ed.), *Socio-organizational factors for safe nuclear operation* (1st ed., pp. 94–97). Montagret: Larsen Science.

Fauquet-Alekhine, Ph., & Labrucherie, M. (2012). Simulation training debriefing as a work activity analysis tool: the case of nuclear reactors pilots and civil aircraft pilots. In Ph. Fauquet-Alekhine (Ed.), *Socio-organizational factors for safe nuclear operation* (1st ed., pp. 79–83). Montagret: Larsen Science.

Klein, D., Simoens, P., & Theurier, J.-P. (2005). Témoignage d'entreprise: une collaboration recherche-industrie conséquente sur l'utilisation pédagogique des simulateurs à EDF. In P. Pastré (Ed.), *Apprendre par la simulation – De l'analyse du travail aux apprentissages professionnels* (pp. 207–220). Toulouse: Octarès.

Le Bellu, S., & Le Blanc, B. (2012). How to characterize professional gestures to operate tacit know-how transfer? *The Electronic Journal of Knowledge Management, 10*(2), 142–153.

Pastré, P. (2005). Apprendre par résolution de problème : le rôle de la simulation. In P. Pastré (Ed.), *Apprendre par la simulation – De l'analyse du travail aux apprentissages professionnels* (pp. 17–40). Toulouse: Octarès.

Proctor, M. D., & Gubler, J. C. (2008). Creating the potential for organizational learning through interactive simulation debriefing sessions. *Performance Improvement Quarterly, 14*(3), 8–19.

The Simulator of Critical Situations in Anesthesia

Thomas Geeraerts and Fabien Trabold

1 Introduction

Risk management in medicine has become a major concern. In France, every year more than 6 million of general anesthesia are performed. Thanks to medical advances and improved care, the risk of death from anesthesia is now very low, about one per million in a patient previously in good health. If anesthesia accidents are rare, they are unacceptable to the public and media attention is often important.

Anesthesia is a risky activity. However, while in commercial aviation, simulators have been developed for a long time, the simulation in anesthesia is still at the beginning. The first realistic simulation model of anesthesia has emerged in the late 1960s. Quickly abandoned because of technical constraints, high fidelity simulation manikins were only available in the 1980s. For structural reasons (high costs, important logistical issue) and due to medical reluctance to use simulation training, simulators are still poorly present in the medical environment. However, most of the critical situations in Anaesthesia and Intensive Care can be reproduced using simulation.

T. Geeraerts (✉)
Department of Anaesthesia and Intensive Care, University Hospital of Toulouse,
Toulouse, France
e-mail: thgeeraerts@hotmail.com

F. Trabold
Haut-Rhin Fire Department, Rescue and Health Service (SDIS 68),
68027 Colmar Cedex, France
e-mail: fabien.trabold@sdis68.fr

F. Trabold
Department of Anesthesiology and Intensive Care Unit, Bicêtre Hospital,
94275 Le Kremlin-Bicêtre, France

© Springer International Publishing Switzerland 2016
Ph. Fauquet-Alekhine and N. Pehuet (eds.), *Simulation Training:
Fundamentals and Applications*, DOI 10.1007/978-3-319-19914-6_4

2 Why Anesthesia is a Risky Situation?

Most of the surgical procedures require general anesthesia with loss of consciousness induced by drugs. This loss of consciousness is accompanied by an alteration of control mechanisms that are fundamental physiological parameters such as blood pressure and respiratory rate. During a deep general anesthesia, there is a significant decrease in respiratory rate to respiratory arrest. If ventilatory function is not supplemented by the use of a artificial respirator, blood oxygenation will deteriorate rapidly leading to cardiac arrest or catastrophic neurological consequences. The role of the anesthesiologist is not only to induce loss of consciousness and analgesia for surgery, but also to implement the means for the replacement of vital functions temporarily impaired. So if respiratory function id not supplemented, such as during a default in the mechanical ventilator, the patient's condition will deteriorate very quickly. Rapid diagnosis of these problems is not always easy because it can be confused with a worsening associated with the patient himself.

General anesthesia is divided into three phases—induction of anesthesia, maintenance and emergence from anesthesia—each with its specific risks: the induction of anesthesia is probably the most risky phase: it consists to induce sleep and to establish, among other things, artificial respiration and monitoring elements. Adverse events, as severe allergy related to drugs or technical difficulties to ensure oxygenation of the patient, can occur during this phase. Maintenance of anesthesia is the period during which the surgery occurs. It can be associated with significant adverse effects such as significant bleeding leading to a decrease in blood pressure associated with hemorrhage (hemorrhagic shock). If blood pressure remains low for too long, organ perfusion will be compromised leading to their irreversible dysfunction and death within days. It is the role of the anesthesiologist to prevent this unfavorable outcome. Finally emergence from anesthesia is the phase where the patient gradually recovers vital functions. This partial recovery is also at risk and warrants specific monitoring.

Anesthesia itself induces risk. However, it is clear that the condition of the patient also importantly influences the risk. If the risk of serious perioperative events is very low for an athlete of 20 years-old undergoing surgery for a broken ankle, it is much more important in a 95 years-old patient undergoing hip surgery. Similarly, the surgery itself significantly affects perioperative risk: Cardiac or thoracic surgery, for example, has a perioperative risk much higher than eye surgery. In medicine, we consider that a risk is important when it exceeds 5 %.

In France, the organization of anesthesia practice is regulated by a strict legal framework. Anesthesia must be performed by physicians specialized in anesthesia and intensive care, which are responsible for the means used and the medical strategy. Anesthesia and intensive care is a medical specialty obtained after six years of general medical training followed by five years of specialized training.

Nurse(s) specialist(s) in anaesthesia and intensive care can help the anesthesiologist. Patients must have a pre-anesthesia visit aiming at evaluating the perioperative risk and defining the anesthesia protocol. It is important to note here that several anesthesia protocols can be used for the same clinical situation. Several and significantly different attitudes are possible and acceptable. The French Society of Anesthesia and Intensive Care (www.sfar.org) regularly publish recommendations for the care of patients with specific conditions during anesthesia.

Due to the improvement of anesthesia security, accidents are now rare (Gaba 2000). Thus, an anesthesiologist will be more and more rarely facing a serious accident. For example, malignant hyperthermia (major dysregulation of body temperature induced by some drugs), a classic complication of anesthesia, will be statistically met once or twice during the career of a practitioner. This complication is rare but very severe, requiring the use of a complex treatment protocol. These accidents are often serious and sometimes unpredictable. In addition, they are rapidly evolving. Response must be given in the minute following cessation of effective ventilation. Irreversible neurological consequences occur after only three minutes of circulatory inefficiency. This time component is crucial to justify the simulator training.

3 Simulators in Anesthesia

The objective of the simulation is to be as close as possible to the reality, allowing the learner and her/his entourage to get involved at best in the simulated situation. The simulator of anesthesia and intensive care may take the form of a mannequin (torso or whole body) or computer software. We will only discuss here realistic simulators. The simulators include a realistic model and a computer interface software allowing the simulation of physiology, pathophysiology and pharmacology. The current models reproduce quite faithfully physiology. Respiratory rate, pulse, heart and lung auscultation are reproduced very realistically. Monitoring devices are also very similar to those used daily in the operating room. The monitor displays basic parameters such as heart rate, electrocardiogram, blood pressure, the percentage of oxygen in the blood or even more complex parameters such as intracardiac pressures (Fig. 1). The implementation of these monitoring elements is flexible. In anesthesia, the decision to set up a complex monitoring (and to use it) is an important component of the strategy to support the patient in a critical situation. It is classic in a patient with high peri-operative risk to implement a greater number of means of monitoring than for patients with very low risk. This decision, which is important for early diagnosis of complications, will be part of the discussion during the debriefing session.

Parameters displayed on the monitor can be controlled by a trainer (who is not necessarily a physician). Some models allow automated responses based on the actions of participants, such as increased blood pressure after adrenaline injection.

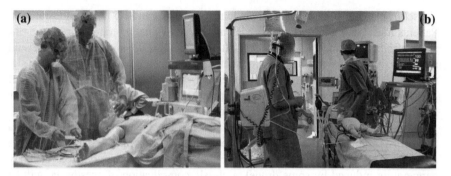

Fig. 1 Simulation session on high fidelity mannequin, adult (**a**) and child (**b**)

4 The Simulation Session

The actors are usually two, simulating the "duo" anesthetist—nurse anesthetist. The roles must be predetermined before the start of the session. The physician's role is to diagnose the cause of the observed anomalies, and to implement appropriate measures. The nurse can help in diagnosis by providing information. His/her technical skills and specialized training should enable rapid implementation of therapeutic procedures that can be very complex. The interactions between the two actors are fundamental. It is a classic example of observing the behavior of deletion of the young doctor in favor of a more experienced nurse, the nurse then taking a leadership role.

The training team is usually composed by experienced anesthetists and nurse anesthetists. Trainers are at least two, usually three, allowing observation of events and reactions occurring in a very short period of time.

The session takes place best in an operating room or in a room resembling as much as possible to the operating room (Fig. 1). The session usually lasts about 30–45 min. It begins with a presentation of the mannequin and equipment by the training teams. The session is filmed and displayed in another room, with the possibly for other participants to watch the session. The clinical simulations are very various, since they can be programmed on a computer (management of cardiac arrest, cardiac rhythm disorders, inability to provide adequate ventilation, a massive hemorrhage or a severe allergy). The simulation session is followed by a debriefing session for 30 min (Fig. 2). This debriefing, led by one or physicians who attended the session simulation, focuses on medical care (which drug injected? what dose? respect of algorithms?), but also on the management and behavioural component of the crisis situation (stress attitude, communication with other). It can draw on the experiences of participants, but also on video recordings of the session. Indeed, the perception of time is often strongly altered during a simulation session. The intensity of events often leads to underestimate the duration of a serious event. A 5 min cardiac arrest with neurological consequences will often be reported as having lasted two to three minutes (usual threshold of severity). Similarly, the

Fig. 2 Debriefing room, which can also be used as a screening room for observers

intensity of stress sometimes leads to surprising reactions (screaming, aggression, or complete inhibition). These reactions are usually completely ignored by the participants, and often a surprise for themselves. Far from wishing to depreciate the participants, the debriefing aims to enable everyone, according to its own means, to improve the management of crisis situations.

5 Applications of Simulation in Anesthesia

5.1 Training of Health Actors

Anesthesia simulators have now reached a level of sophistication allowing their use as a tool for initial and continuing training. They can provide group training in crisis management (Gaba 1992).

Patient simulator is a tool of the undergraduate medical education. The use of a simulator in training has several advantages for the learner as for the patient. For the learner, s/he is spared the "stress of the first time": the initial training on the mannequin to repeat the same movement, the same situation, until a perfect control, is safe for the patient. This pedagogy using repetition is not possible on a patient. The simulator also allows the learning of new technology.

The objectives of the use of simulators for the qualified practitioner and continuing medical education may be different. For a qualified practitioner, the simulator can test and improve the interaction between the practitioner and the team in managing a crisis. The simulator allows training in critical incident management with consideration of human factors in the cascade of events. The objectives of the training simulator are then (1) early recognition of the incident, (2) calling for reinforcements, (3) taking the lead in actions, (4) managing available resources and (5) allocating work tasks with communication between different actors. For example, the repetition of an algorithm for management of cardiac arrest or severe trauma scenario allows the team to test how it works together. Moreover, the use of a simulator allows for standardization of care between different teams. Like the benefit of the simulation as part of initial training, it has been shown that simulation can significantly improve the performance of individuals facing a critical situation (Chopra et al. 1994).

5.2 The Simulator as a Tool for Clinical Research

The value of the simulator is to reproduce identical clinical situations. In particular, it is possible to simulate difficult intubation conditions (introduction of a tube into the trachea to ensure adequate ventilation). Twigg has shown that intubating conditions were worse with blades of disposable laryngoscope than with reusable ones (Twigg et al. 2003). With the simulator, intubating conditions were fully comparable between each attempt at intubation. A recent study showed that in a crisis simulator, errors in drug dosages are directly dependent on the spelling of the dosage (inscription easy to read in milligrams per milliliter or more difficult as a percentage requiring conversion) (Wheeler et al. 2008).

6 Limits of Simulators in Anesthesia and Intensive Care

6.1 Financial and Technical Issues

The direct costs of simulator are important, ranging from 50,000 to more than 150,000 euros, and maintenance costs must be also considered. In addition, the scenario is more credible if the whole environment is close to reality. Thus, it is important to recreate an operating room or a pre-hospital ambulance resuscitation in order to immerse the learner in the situation.

In addition, personnel costs are very high. They include training instructors (anesthetists and nurse anesthetists) and the cost of hours spent in instruction. These high costs can be an obstacle to the development of the simulation. Indeed, in medicine, unlike other areas, the costs of simulation are higher than the actual situation (i.e. one hour of flight time).

Although simulators are improving, all clinical situations can not be perfectly simulated (including obstetrics). In addition, the mannequin still lacks realism. The skin changes of great importance in medicine are still sadly lacking and greatly reduce the realism of the mannequin.

6.2   Unresolved Issues

Our experience from several years now has allowed to raise important issues.

What if inadequate care? Should we go to the ultimate evolution of the scenario and the patient's death if care is considered to be not adequate? To our knowledge, there is no assessment of the consequences of this attitude on the learning of critical situations in medicine. The patient's death during a scenario could lead to defensive and counterproductive reactions by participants, as pointing out the lack of realism of the mannequin or the lack of credibility of the situation. Again, the contribution of the recording session and viewing by the participants could help.

The evaluation of practitioners in the simulation remains a challenge. For us, the use of simulation in anesthesia is limited to training. Physicians (and authority) are still far to consider a declaration of incapacity followed by retraining of a practitioner who failed a simulation session, as this can be done in aviation. The very hierarchical nature of the medical profession makes it difficult to be assessed by colleagues of the same "level".

Finally, if the algorithms have been developed to optimize the management of critical situations, it happens that significantly different management can be successful in terms of patient recovery. In an educational point of view, the respect of procedure and algorithms are fundamental; however, the goal of the medical profession is mainly patient's recovery. The complexity of some situations makes them difficult to standardize. The interaction with the trainers, who can change the response of the mannequin based on the actions of the participants, becomes here essential. Even if the management of the situation significantly differs from the algorithm, it can be conceivable to positively change the evolution of the scenario, if the management was finally correct. The prerequisite for this is a good debriefing emphasizing the transgression of the rules, but respecting the success of the treatment. Pedagogy using simulator finds here his human side, essential for learning in Medicine.

7   Conclusion

Improving security in anesthesia led to a drastic reduction in accidents. Anesthetists are more and more rarely facing a serious accident. Therefore, learning only based on experience does not sufficiently train. Moreover, in anesthesia, appropriate responses should be implemented very quickly (within a minute). High fidelity simulation allows learning to manage these critical situations. This learning will probably become mandatory in the coming years as part of continuing medical education. The future is likely to be trained in a multidisciplinary approach combining for example in the operating room, the anesthesia team and the surgical team.

Simulation is an educational tool in complement to books teaching and clinical experience. Work on simulator will not replace other teaching but will improve the quality of learning to further reduce mortality in anesthesia.

Insert

Anaesthesia in Intensive Care
   The simulation can improve the performance of the team facing a crisis, and develop

- Personal Resources:

  - Observe, cogitate, adapt,
  - Initiate treatment,
  - Re-evaluate, prioritize,
  - Declare.

- Resources Group:

  - Leadership,
  - Communicate, delegate, cooperate, search for help and specific skills.

References

Chopra, V., Gesink, B. J., de Jong, J., Bovill, J. G., Spierdijk, J., & Brand, R. (1994). Does training on year anaesthesia simulator lead to improvement in performance? *British Journal of Anaesthesia, 73*(3), 293–297.

Gaba, D. M. (1992). Improving anesthesiologists' performance by simulating reality. *Anesthesiology, 76*(4), 491–494.

Gaba, D. M. (2000). Anaesthesiology as a model for patient safety in health care. *BMJ, 320*(7237), 785–788.

Twigg, S. J., McCormick, B., & Cook, T. M. (2003). Randomized evaluation of the performance of single-use laryngoscopes in simulated easy and intubation difficulty. *British Journal of Anaesthesia, 90*(1), 8–13.

Wheeler, D. W., Carter, J. J., Murray, L. J., Degnan, B. A., Dunling, C. P., Salvador, R., et al. (2008). The effect of drug concentration expression we epinephrine dosing errors: A randomized trial. *Annals of Internal Medicine, 148*(1), 11–14.

Virtual Surgical Simulation: The First Steps in a New Training

Luc Soler and Jacques Marescaux

1 Introduction

Laparoscopy, in other words the introduction of an optical device into a patient's abdomen, which allows for realizing the surgical manoeuvre thanks to a miniaturized camera, represented the most important change the surgical world experienced in the 20th century: the "minimally invasive" surgery area was born. This surgery has many advantages for patients by strongly reducing the risks of nosocomial infections,[1] as well as shortening the hospital stay or reducing postoperative pain. Conversely, it represents a greater surgical complexity than conventional surgery and requires specific, sometimes time-consuming training in the case of the most complex procedures. Indeed, in order to operate, surgeons introduce a long optical device connected to a camera into the patient thanks to a small incision with a 5 mm to 1 cm diameter. Thus they see the inside of patients as well as their gestures on a screen with greater accuracy. However, what they see is in two dimensions, unlike natural stereoscopic vision[2] during open surgery. Furthermore,

---

[1]Nosocomial infection: infection induced by a patient's hospital stay.
[2]Stereoscopic vision: Vision induced by the differing perception between the right and the left eye of a same scene. The difference of both views enables the brain to reproduce the perception of relief.

L. Soler (✉) · J. Marescaux
IRCAD Institute, University of Strasbourg, IRCAD, 1, place de l'Hôpital, 67091 Strasbourg Cedex, France
e-mail: luc.soler@ircad.fr

J. Marescaux
e-mail: jacques.marescaux@ircad.fr

© Springer International Publishing Switzerland 2016
Ph. Fauquet-Alekhine and N. Pehuet (eds.), *Simulation Training: Fundamentals and Applications*, DOI 10.1007/978-3-319-19914-6_5

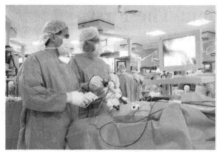

Fig. 1 Comparison between real minimally invasive surgery (*left*) and its virtual simulation (*right*): interfaces and the operative mode enabling practitioners to carry out surgical manoeuvres are identical

to operate, they use long instruments that are introduced into patients through small incisions of 2–5 mm. They lose then part of their haptic perception,[3] also called force feedback. Indeed, only kinaesthetic perceptions can be preserved during tool contact with an organ, tactile perception used to palpate organs or tactile search for tumours in open surgery are no longer possible. These new difficulties however offer an important advantage: surgery can be simulated virtually. Indeed, the real patient image can be replaced by a virtual image and the long instruments can be connected to force feedback systems, thus accurately reproducing the real interface of surgery between surgeons and their patient (Fig. 1).

In the 1990s, the only simulation that could be achieved consisted in using a box in which objects were inserted that had to be manipulated with the real instruments of minimally invasive surgery. These "endo-trainers" were progressively perfected and led to real human dummies featuring plastic organs that could bleed (e.g. Simulab Corporation, Limbs and Things). The main drawback of these training systems is the high cost of artificial organs that have to be replaced after each training session, while the realism remains very questionable compared to a real patient. Maximal realism is obtained when operating animals, as in the greatest training centres such as the European Institute of TeleSurgery of IRCAD (Institut de Recherche contre les Cancers de l'Appareil Digestif). One of the drawbacks of that kind of training, just like the endo-trainer, remains the subjective aspect of the evaluation, the quantifiable data resulting from such training being very limited. Furthermore, simulated pathologies are not present in the animal. Finally, the anatomy of the porcine model, though similar to human anatomy, is not identical. It is however currently the best training method, the closest to reality since the living animal is set up under general anaesthesia in the same conditions as a patient.

Unlike these "real" simulations, virtual simulation provides multiple quantifiable and exploitable data in order to objectively evaluate students. Over the last 5 years,

---

[3]Haptic perception: perception linked to the sense of touch (tactile) and the mechanical reaction of the body's muscles (kinaesthetic).

many virtual surgical simulators have thus been developed. They rely on the virtual modelling of the human body combining a geometrical and a mechanical model. The interface with surgeons is represented by dummy surgical instruments which are connected, in the most advanced solutions, to force feedback systems allowing to reproduce the kinaesthetic perception induced by the contact of the instrument with virtual organs. Thus, force feedback is an essential element for any realistic surgical simulator. Today, there are many haptic systems for the reproduction of free manoeuvres of instruments such as the scalpel or the needle (Phantoms from Sensable or Omega systems from Force Dimension) or constrained and limited manoeuvres of instruments such as laparoscopic tools[4] Laparoscopic Impulse Engine from Immersion, Instrument Haptic Port from Xitact, Force Feedback System from Karl Storz). The most simulated minimally invasive surgery procedures are digestive laparoscopy, urology, gynaecology and arthroscopic procedures. Several companies currently propose such simulators, among which Surgical Science (www.surgical-science.com), Simbionix[TM] (www.simbionix.com), SimSurgery (www.simsurgery.no), CAE Healthcare (www.caehealthcare.com) and Mimic Simulation (www.mimicsimulation.com). These simulators all propose simulation scenarios under thematic modules. These fairly recent simulators have now acquired some maturity thanks to the use of realistic rendering both on visual and haptic level. They allow thus to learn basic manoeuvres in minimally invasive surgery in a step by step mode, leading in the end to more complex surgical manoeuvres. It is finally important to point out that all these systems have been developed with the aim of autonomous training without systematic supervision by a teacher. They feature an automatic marking system which logically has to enable students to learn on their own. This is certainly one of the major differences with professional simulators described in this book and used by companies in the fields of aeronautics, energy or iron and steel, which are all used under the supervision of an expert. This leads to one of the major questions those simulators raise today: is an efficient teaching via virtual simulation possible without teacher or without expert?

Literature is full of validations showing the benefit those virtual simulators provide (Tjiam et al. 2014; Shaharan et al. 2014; Amirian et al. 2014). The major one is gaining time in the training on the living. However, we have noticed a problem which is recurrent in virtual simulation. Three of the main virtual simulators currently available on the market have been tested in their "basic manoeuvre" configuration both by expert surgeons in minimally invasive surgery at IRCAD and by engineers in computer sciences specialized in virtual simulation. These basic manoeuvres allow for the training of students in camera and surgical tool manipulation. It also teaches them to coordinate their manoeuvres. An expert surgeon should logically carry out that kind of manoeuvre easily and get the highest marking from the system. This is not what happened in reality since computer engineers systematically had better marks during the first manipulations. In fact, by

---

[4]Laparoscopy: surgery consisting in using long instruments introduced inside the abdomen of a patient through small orifices of 2 mm to 1 cm diameter.

accurately studying those sessions, we notice that expert surgeons needed several sessions before perfectly mastering the system. They translated that adaptation time themselves as learning time of the system, the lack of realism of the scene being the main drawback for surgeons, who could not locate their natural landmarks. This shows a deep problem. Although current virtual simulators provide a benefit in training, they do not necessarily preserve expertise. The benefit is hence open to criticism and the use of those systems could therefore be questioned.

In order to get around those problems, one of the solutions consisted in developing "patient-specific" simulators, i.e. incorporating photo-realistic rendering and accurate modelling of patient anatomy from 3D medical imaging. These simulators were up to then limited to research projects, but have recently been commercialized by young companies (Digital Trainers in France, or VirtaMEd in Switzerland). These highly realistic simulators offer the advantage of preserving expertise while maintaining the benefits of more traditional virtual simulators. Our study will rely on two simulators of that kind, the ultrasound guided[5] simulator "HORUS" and the laparoscopic surgery simulator ULIS, both developed by IRCAD (Forest et al. 2007; Soler et al. 2008; Nicolau et al. 2011; Hostettler et al. 2013).

2 Simulation Situation

Our study has been realized on the basis of two "patient-specific" simulators allowing to learn and train various procedures. In both cases, the students are medical students (see Insert 1) who are carrying out a non-residential internship (5th year of medical studies after A level examination) for one month within the Research and Development department in computer sciences at IRCAD. This internship takes place either before or after an internship in the radiology department. The objectives of the training are:

- Learning how to handle an ultrasound (US) probe, including mastering the probe positioning, reading the associated image and spotting the visualized structures in space.
- Learning the manoeuvres of needle insertion under ultrasonic control, in particular for hepatic biopsy[6] and liver tumour destruction (called hepatic tumours) through thermal ablation.[7] This includes mastering the coordination of manoeuvres between the ultrasonic probe, which is held by one of the hands of the operator, and needle positioning, which is held by the other hand. An identical coordination of manoeuvres is required for any intervention consisting in inserting a needle in various anatomical regions under ultrasonic control.

---

[5]Ultrasound guided: gestures guided by ultrasound such as the introduction of a needle inside the body.

[6]Hepatic biopsy: retrieving liver (hepatic) cells by core drilling with a hollow needle.

[7]Thermal ablation: thermal destruction with a needle conducting or creating heat or cold at its tip.

- Learning basic manoeuvres in laparoscopic surgery including mastering the camera positioning, mastering manoeuvre coordination between several instruments and mastering suction tools allowing for blood suction during haemorrhage and electrocoagulation tools used to burn tissues and coagulate blood with the electric arch of a specific tool.

Insert 1

In France, medical studies are divided into three cycles. The first cycle of medical studies (PCEM in French) is done during the first two years. The first year (PCEM1) allows for a selection through a competitive examination offering a limited number of available places (numerus clausus) for the second year (PCEM2). This first cycle essentially provides theoretical teaching, a 3–4 week clinical internship done between both years being the only practical immersion into the clinical environment. The second cycle of medical studies (DCEM in French) is done from the third to the sixth year. The third year (DCEM1) represents the transition from theoretical training to a more pathology oriented training. Students learn during this year to carry out the clinical examination of a patient. The fourth (DCEM2), fifth (DCEM3) and sixth (DCEM4) years are called "non-residential" because of the former competitive examination to enter French non-resident studies which has been abandoned in 1968. During those three years, students make one to three month internships in hospitals in various departments of the university hospital. At the end of the four years of the second cycle, students take the national ranking exam (ECN) which replaced the former competitive examination for non-resident studies. It allows to rank students who will choose their six month hospital internships to be done during the third cycle. The third cycle of medical studies (TCEM), where students are called interns, lasts three years for general medicine, and a minimum of four years for other specialities. During those internships, students have the opportunity to get some medical practice under the supervision of experimented practitioners. Students will perform surgical manoeuvres when they reach the resident level.

2.1 HORUS: Simulation of US-Guided Procedures

Haptic Operative Realistic Ultrasonic Simulator (HORUS) is a simulator allowing to virtually reproduce an ultrasonic image from a CT scan image (Computer Tomography) or an Magnetic Resonance Imaging (MRI). It also allows to simulate the insertion of a needle inside a patient's body with force feedback in order to carry out two surgical manoeuvres: biopsy and thermal ablation (Fig. 2).

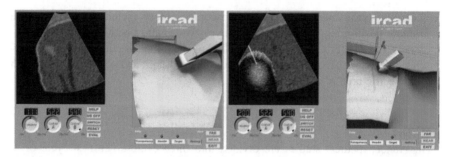

Fig. 2 Two views of the HORUS simulation software illustrating the simulation of US images and the simulation of thermal ablation after needle positioning in the virtual patient

In its educational version, the software proposes to do that kind of manoeuvre on a clinical base of five different patients suffering from liver tumours (Fig. 3). When a student selects one of the cases, the surgical procedure to be performed is described by the software. The student then enters the simulation mode so as to control the US probe with one hand and the needle with the other, which will have to be inserted into the targeted tumour. Using a real US probe that has been cleared from its electronic content enables greater immersion into the simulation. We have however not developed a system based on a real biopsy or thermal ablation needle, the needle being represented by the pen of the Omni system. Indeed, these two elements are physically connected to two Omni-type force feedback systems from Sensable. This allows to reproduce the kinaesthetic feedback induced on the one hand by the contact between the US probe and skin, and on the other hand by the needle piercing the skin and underlying tissues (Fig. 3).

The simulator offers a certain number of functionalities (see Insert 2) among which one of the most important ones is automatic evaluation. The system provides essential quantitative and qualitative information: the percentage of pathological tissue destructed in case of thermal ablation and the percentage of pathological tissue retrieved in case of biopsy. This information also indicates if the student went through tissues or not which did not have to be touched during the procedures as well as the position of the needle with respect to the target. It does however not give a mark. Being experimental, no rule of usage of the simulator is given. It can be

Fig. 3 Illustration of the educational version of HORUS for the simulation of US procedures from two Omni-type force feedback systems from ©Sensable held by the student (*right*)

used autonomously without any supervision by a teacher or in a supervised fashion. Both usages will be analysed in order to extract from each teaching mode the advantages and drawbacks, as well as basic rules of good practice.

Insert 2

During simulation, users can choose a certain number of settings, which are also featured on real ultrasonic equipment, for instance the frequency and image contrast can be set. During each exercise, four rendering modes that can be modified during the exercise are proposed to users. The "transparency" mode enables to see patients in transparency as well as the plane of the ultrasonic scanning equipment in 3D. Students use this mode when beginning their training. The "needle" mode permits to see the optimum position of the needle in transparency, as it should be positioned to reach the aimed tumour. The "target" mode makes the position of the needle entry point appear on the skin. Finally, the "nothing" mode provides no additional visual support (Fig. 2). Confirmed users or experts use this training mode. All four modes are used progressively by students, as they advance in their training. Finally, students can at any moment request support from the software so as to remind them of their aim. They can also request an automatic evaluation of their procedure.

2.2 ULIS: Laparoscopic Surgery Simulation

Unlimited Laparoscopic Immersive Simulator (ULIS) is the first patient-specific laparoscopic surgery simulator using the new force feedback system developed by the Karl Storz company. This system is composed of two or three entry ports with force feedback allowing for the insertion of real laparoscopic instruments that have been adapted for the system (Fig. 4). Unlike conventional simulators using virtual

Fig. 4 ULIS uses real surgical instruments to interact with haptic feedback on the virtual patient of the simulator, making it more realistic

Fig. 5 Exercise examples of the ULSI simulator: camera manipulation (*left*), haemorrhage suction (*centre*) and electrocoagulation with smoke effect (*right*)

tools that are often little resembling and that in particular do neither have the same dimensions nor the same operating radius. Using real tools allows for a more realistic teaching and greater immersion.

Furthermore, the simulator includes a database of clinical cases that have been modelled in three dimensions from their CT images (Soler et al. 2007), and then textured photo-realistically for a rendering close to reality (Fig. 5). This simulator offers the opportunity of progressive training relying on six basic exercises to learn the basic rules of laparoscopic surgery (see Insert 3).

The simulator features a student management mode in order to record all the automatic evaluations provided by the simulator at the end of each exercise that has been done. Just like the HORUS simulator, this experimental simulator has no specific rule of usage. It can be used in an autonomous manner without teacher supervision or in a supervised manner. We will later analyse both manners in order to extract the advantages and drawbacks from each teaching mode, as well as basic rules of good practice.

Insert 3
Users first have to learn camera manipulation alone. Then, they learn to manipulate two surgical tools. The following exercise consists in coordinating their movements between a camera on the one hand and a tool on the other hand. They learn then to coordinate their manoeuvres between two tools which have to interact simultaneously. The two last exercises are to learn how to aspirate blood coming from a haemorrhage and to coagulate points located inside the abdominal cavity, while controlling the tool with one hand and the camera with the other hand. Each exercise features a set of options, among which the inversion left hand-right hand (surgeons have to master their instruments with both hands) and the angulation of the field of view of the camera (30° cameras being more difficult to master).

2.3 Simulation Conditions

In order to establish optimal rules of usage for simulators, we have tested them under various configurations of use. One of the main objectives was to check the opportunity to realize non supervised training by using the automatic evaluation of simulators. Indeed, current teaching of surgery is done through companioning. This means that a teaching surgeon teaches students the daily hospital practice. Practical teaching of manoeuvres is done during surgical interventions on patients. A comprehensive teaching of surgical procedures beside that clinical time would thus require allotting more time to teaching practitioners, which is not possible due to the lack of surgeons. Specific teaching, beyond the clinical environment, where the teacher fully dedicates to training beside the hospital practice is therefore limited to a few centres such as IRCAD. That kind of teaching, currently being carried out on animals, is quite expensive, all the more since it is difficult for surgeon teachers to have time for such training. One of the hopes induced by virtual simulation relies thus on computer-assisted teaching, where the surgeon teacher would not have to follow students throughout the entire virtual intervention. This would therefore permit to pursue the daily hospital practice while in parallel students train on simulators. Is this possible and under what conditions? Is there a minimum acceptable degree of involvement from the teacher and if yes, what is that degree? Or has the teacher to be present throughout the entire simulation, like instructors on flight simulators? We propose to answer those questions with various teaching experiences.

3 Analysis of Training Through Simulation

In this chapter, we will analyze training provided by both previously described simulators HORUS and ULIS following two major modalities: autonomous learning and learning with companioning. For each example, we will produce remarks which may become rules of usage of these simulators. We will then be able to extrapolate these specific rules so as to set up more general recommendations of an optimized use of simulation for learning.

3.1 Learning US-Guided Manoeuvres: HORUS

We use HORUS since November 2005. This limited and experimental teaching does of course not propose the same level of expertise as flight simulators which have been used in routine for many years. It nevertheless enabled us to compare two very different teaching modes on this simulator: the autonomous mode and the accompanied version. They both present advantages and drawbacks that we will highlight.

3.1.1 Autonomous Learning with HORUS

- **Familiarization with the simulator by the student: presentation of the simulator**

Whatever mode used, familiarization with the simulator requires an initial presentation step of functionalities and description of how HORUS works. That initial step, companioning type, remains limited to 30 min in the case of autonomous learning. It could be done by a technician who would merely present functionalities. In the case of HORUS however, which does not feature an evaluation system for US probe manipulation, this would lead to an immediate problem: students would have neither recommendations nor understanding of the guidance tool. We have noticed this quite rapidly during the first manipulations and it showed us a limit of the evaluation system of the HORUS simulator in the autonomous mode. As we will later confirm it, this induces a first development rule of that kind of simulator.

Rule 1: autonomous learning on a simulator requires an automated evaluation of students by the simulator

The use of the US probe not being evaluated by that simulator, the initial simulator familiarization through a companioning phase is hence mainly oriented towards the use of the US probe. During this phase, students can ask as many questions as they want to the teacher. The thirty minutes are split up as follows:

- 10 min software (launch, options accessible from the interface, explanation of the views of the interface) and force feedback system presentation (two Omni systems from Sensable, one for connection and control of a US probe and the other one for control of the virtual needle).
- 15 min demonstration of an exploration simulation[8] in order to explain how to use the US probe and how to spot the anatomical and pathological structures inside these images according to the orientation of the probe.
- 5 min demonstration of a simulation so as to remind simple rules of needle insertion inside the body to perform a biopsy or a thermal ablation (preserving vessels, limiting the number of pierced organs, minimizing risks) and mainly to explain the use of ultrasonography to guide the manoeuvre (the plane of the US image has to contain the needle so that it is visible in the image).

This phase ends with an advice on how to use the simulator progressively, which represents an educational approach. This proposed approach is all the more important since the software will be used in autonomous mode. The student can hence use the simulator in a non-reasoned manner. The description of an educational approach provides a framework to training and enables to optimize it. The teacher thus proposes to start training by exploring five accessible anatomical

---

[8]Exploration simulation: exploration of an anatomical cavity through a visualization tool; US probe or camera.

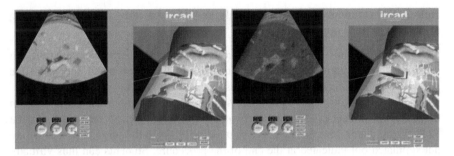

Fig. 6 The transparency mode helps locating the visualized structures in the native CT image (*left*) or on the virtual US image (*right*)

cavities, starting each with visualization in scanner mode (the native image is not modified) with transparency, then switching to the US image simulation mode, ending without patient transparency (Fig. 6). Once the exploration of the cavity is mastered, the teacher then advises the student to continue the training with biopsy and thermal ablation manoeuvres. This progression allows to start in a mode where the visualized image is easier to read (the CT image has no shadows unlike the US image) with the visualization of the US plane in transparency inside the virtual patient. This helps understanding the cause-effect relation between the position of the US probe on the abdomen and the visualized image. It further allows to locate visualized organs thanks to the virtual transparency of the patient.

- **Use of the simulator by the student**

In this autonomous mode, at the end of the familiarization phase, students are free to use the simulator as they want. They have five clinical cases with greatly varying pathology localizations and also varying reading difficulties of the US image between patients. First, we notice that this free usage prejudices training. Indeed, most students do not follow the recommendations for use of the simulator provided by the teacher during the familiarization phase. Thus, the most standard use of the simulator in that mode complies with following pattern: selection of a patient and rapidly trying out the biopsy or thermal ablation manoeuvre. The prior exploration phase is then reduced to a minimum, students using the transparency mode (Fig. 6) to facilitate the guiding. After about 5 h of simulation, students however realize that they are not able to carry out the manoeuvres without the transparency mode and spend then much more time on exploration. They go back to the proposed educational approach rather naturally. In the end, the freedom of manipulation logically entails losing time when learning on a simulator since its use is in this case not optimal, no educational approach being provided. This evidence leads to a second rule of autonomous use of a simulator.

Rule 2: the autonomy of use of a simulator has to be compatible with sticking to an educational approach

This rule is well known in the world of computer games, which propose increasing difficulties in the game, access to certain parts of the game being subject to the successful undertaking of previous "missions".

Another important point regarding this autonomous training mode is the freedom of action of students, who can carry out the entire surgical procedure. In the HORUS simulator, the consequences of possible errors in the surgical procedure are not simulated but indicated at the end of the procedure. Students can thus virtually carry out serious mistakes that could lead to important postoperative consequences ranging from the necessity of a new surgical intervention to the patient's death. Whatever error, it will only be indicated by textual information on the direct consequences of the procedure: patient haemorrhage pierced healthy organs, etc. This freedom of action with consequences limited to merely written information induces quite some lack of concern from students towards the consequences of their errors. In the reality of training as it is currently taking place in clinical routine on real patients under the control of the teacher, students could never carry out such a manoeuvre, putting patients' lives at risk. Students therefore rapidly make the difference between simulation, where they make sometimes dangerous manoeuvres, and reality, where they would be much more cautious and would evidently limit taking risks in terms of chosen approaches. It both has an advantage and a draw-back. The advantage lies in the fact that students can learn from their errors and can try out complex manoeuvres they would otherwise not be able to test in reality. The drawback is the absence of mental immersion in the performed surgical procedure, leading to sometimes useless risks for patients. As we will see it later on, these benefits can be exploited favourably while limiting the drawbacks by adding a teacher behind students. This takes students back to the training mode they would have known with a real patient, what makes the virtual patient more real.

Another way of solving the mental immersion problem would consist in simu-lating the immediate negative consequences of an error, in particular by simulating continuous tension monitoring of the patient (including a sound signal leading to real stress in students when exceeding normal limits), an ECG[9] (noisy environment adding significant stress in case of heart failure) or haemorrhages (dropping blood pressure and bleeding). Adding a video showing a patient being taken to the sur-gical ER and undergoing surgery for the damaged organ could also make the system more complete and illustrate postoperative consequences of an error. In fact, all these systems would increase realism and raise student concern towards their errors, which are linked to their lack of perception of the consequences their virtual manoeuvres may have. From this we deduce an additional rule close to the first rule but adding the concept of postoperative consequences of manoeuvres.

---

[9]Electro-Cardiogram: system used to monitor a patient's heart rate.

Rule 3: the autonomy of use of a simulator requires automatic and immediate feedback on the efficiency of a manoeuvre and its consequences

- **Evaluation of students**

We chose to limit autonomous training to a total duration of 20 h of use. The evaluation cannot be compared to an exam. It has been set up to assess the efficiency of the simulator to learn ultrasonography and needle placement under US control. This evaluation helped us in particular noticing the effects induced by the differences in learning between the autonomous and the accompanied learning mode. Since the training is done on images of patients suffering from liver tumours, optimal evaluation would have consisted in making students carry out a hepatic biopsy on a real patient. In terms of ethics this is of course absolutely not conceivable. Patient anatomy being furthermore greatly variable, a same operative difficulty for all students could not be guaranteed. This last limit would be present if we would have used an evaluation on animal, i.e. on simulator with living organs. We must not forget the prohibitive cost of such an evaluation which could only be justified for the sporadic evaluation of educational efficiency of the simulator on a limited number of students. Another possible evaluation mode would consist in having an *"internal validation"*, where it would be checked on the simulator whether a trained task has been acquired. This validation has the advantage of using the simulator used for training for evaluation purposes. It simplifies thus the development or evaluation work. Conversely, the drawback is that it is not possible to make the difference between students who are able to reproduce a learnt manoeuvre on a set model on which they trained and students who are able to reproduce that same learnt manoeuvre on a new patient suffering from a comparable pathology. As we have developed a comparable simulator but dedicated to training other manoeuvres, we opted for an *"external validation"* which consists in evaluating the learning of concepts of a domain on an operative mode that differs from the one used. Students are therefore facing situations during the evaluation on simulator that they did not encounter during the training phases. In our case, evaluation consists in performing US exploration of the abdominal cavity of a pregnant woman on a simulator, then in performing an amniocentesis[10] (Fig. 7). Images as well as manoeuvres are comparable but still different. They nevertheless allow to efficiently control if students master the manipulation of a US probe, are able to locate anatomical structures on US images, also if they can manipulate and coordinate their manoeuvres during needle positioning and insertion.

Thanks to these evaluations, we notice a significant variation between students in terms of efficiency of the needle insertion manoeuvre and the US probe manipulation during the exploration phase. Our experience shows that students who follow the proposed educational approach are more effective, essentially because they better control the US probe manipulation, which is fundamental both for locating a

---

[10]Amniocentesis: using a needle to take amniotic fluid, which is in the sac around the fœtus protecting it.

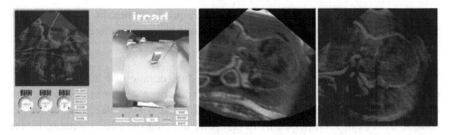

Fig. 7 Simulation of US-guided amniocentesis on a pregnant woman (*left*), MRI of the patient (*centre*) allowing to simulate a US image (*right*)

tumour and for reaching it with the needle. Therefore it confirms the aforementioned rule 2. For whatever student, it however shows that control is not optimal. One of the explanations to this limitation is certainly the absence of companioning. Indeed, due to the fact that the software does not provide any evaluation for US probe manipulation, it is difficult for students to set up efficient probe manipulation rules. As we will see later, this difficulty can be overcome thanks to the presence of a teacher who explains how to manipulate the US probe during the virtual intervention. Here the aforementioned rule 1 kicks in, i.e. autonomous learning on a simulator requires an automatic evaluation of students by the simulator.

3.1.2   Accompanied Learning on HORUS

- **Familiarization with the simulator: presentation of the simulator**

As for autonomous alearning, familiarization with the simulator requires an initial presentation step of functionalities and a description of how HORUS works. This initial step takes an hour, which is twice as long as for autonomous learning. It can be divided into three steps, similar to the previously seen steps in the autonomous learning mode. During these steps, students can ask as many questions as they want:

- 10 min software and force feedback system presentation.
- 25 min demonstration of an exploration simulation in order to explain how to use the US probe and how to spot the anatomical and pathological structures inside these images according to the orientation of the probe. Simulation is done on a minimum of two patients.
- 25 min demonstration of a simulation so as to remind and illustrate simple rules of needle insertion inside the body to perform a biopsy or a thermal ablation and mainly to explain the use of ultrasonography to guide the manoeuvre. Simulation is done on a minimum of two patients in order to carry out a biopsy and a thermal ablation so as to differentiate the result of each procedure, especially in terms of automatic evaluation.

- **Use of the simulator by students**

After the familiarization step, the teacher details the educational approach students will have to follow. This approach is composed of several steps enabling students to progressively master all the elements of the taught manoeuvre. We are here at step 1 of simulation according to the aeronautics terminology.

Step 1 consists in starting the training by exploring five accessible anatomical cavities, beginning for each one with the visualization in CT mode (the native image is not modified) with transparency (Fig. 6) and systematically comparing these images to the simulation mode of US images. Exploration is done by students accompanied by the teacher who controls the switch from CT to US view. This way students see the difference between a view that is easier to read and generally well known at this state of medical education, the CT image, and a little or even not mastered view at all, the US image. The teacher can take control of the US probe if need be and is present to guide students and explain them how to discover structures of interest on each of the five clinical cases (15 min for the first and second clinical case, 10 min for each of the other three). The teacher also indicates how to position the probe, thus visualizing the standard anatomical planes: frontal, sagittal and axial, as illustrated on Fig. 8. The teacher does however not show students how to place the needle. This will be done at a later stage.

In step 2 students navigate alone inside all five abdominal cavities. This "free flight" step is done autonomously by students without the control of the teacher and during five hours. After those five hours, the teacher joins the students who will explore the various abdominal cavities in US mode without skin transparency during one hour. Each organ, each anatomical or pathological structure is then visualized under teacher control, who can intervene at any time to help students so as to improve their manoeuvre. The teacher can then switch to the transparency view of the patient so that students better apprehend the visualized structure. It is thus not an exam but a check of the US probe manipulation control level for exploration. We have noticed that at this stage of training, i.e. 7 h after the beginning of the training, most students pretty well master the US probe and perform a good exploration of the cavity. Compared with the evaluation done after 20 h of autonomous simulation, the quality is even higher. This would therefore tend to show that training with companioning on HORUS is more efficient for that manoeuvre than autonomous training. This fact has however to be relativized since

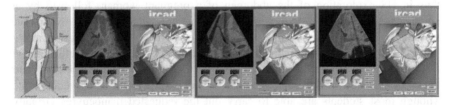

Fig. 8 The three anatomical cutting planes, axial (*yellow*), frontal (*blue*) and sagittal (*red*), illustrated on the left can be found with a US probe

there is no automatic marking system that could on the one hand objectively evaluate this difference, and on the other hand improve autonomous training thanks to a system that would enable students to know if they are making good or bad manoeuvres. It is therefore not a general rule but a HORUS-own fact.

As for steps 1 and 2, steps 3 and 4 are executed to teach and then practice needle manipulation and insertion under US control. For an hour students perform three biopsies and two thermal ablations in step 3 proposed by the simulator under the supervision and thanks to the advice of the teacher. They will autonomously reproduce the manoeuvres taught during 5 h in step 4, followed by teacher control during one hour. This allows to correct the manoeuvre and coordination errors or imperfections. Indeed, unlike exploration, this manoeuvre is harder to master by students. Thus, after a specific 6 h training dedicated to needle insertion, students do not sufficiently master the insertion. There is a manoeuvre evaluation system for these exercises. Once the manoeuvre has been performed, it indicates whether they have pierced vessels or organs they were not meant to touch and if they have reached the aimed goal (percentage of pathological cells taken for biopsy and percentage of destroyed pathological tissue for thermal ablation). There is also a virtual guide, indicating them the optimal needle position so as to reach the targeted tumour. We have noticed that in autonomous mode, students very often use the guide during the five hours. As they have trained with a guide, it will obviously be more difficult during the exam with the teacher, since they can no longer use the guide. Once again, this shows the limits of our simulator in autonomous mode. Indeed, it offers support opportunities to explain students how to do a perfect manoeuvre, but it can conversely not guarantee the judicious and educational use of that support. Here again, we must not rush to the conclusion that an autonomous system cannot be efficient. We can indeed imagine that it could be possible to add a scenario with a sliding scale of guide use. Solutions are thus probably possible, but they are more difficult to realize and sometimes even to find. It seems on the contrary all the more evident that companioning is a solution to guarantee an efficient exploitation of the simulator.

After the exam, students undergo another 5-h autonomous training session during which they are not allowed to use the guide. After that guided 20-h training, students end their training with an evaluation phase.

• **Student evaluation**

The evaluation is the same as the one described for autonomous training. It consists thus in US exploration of the abdomen of a pregnant woman, followed by an amniocentesis done with the same HORUS simulator. Thanks to these evaluations we could notice that all students well master exploration under ultrasonography. The HORUS simulator in companioning mode is thus a particularly efficient tool for this teaching. However, compared with US exploration, needle insertion is not that well mastered by students, although it is way superior in autonomous mode. Though most students are able to carry out the requested manoeuvres virtually, some of them are unable to do them efficiently after 20 h of simulation, 12 of which are specifically dedicated to that manoeuvre. This leads to three hypotheses: the first one is that HORUS does not allow to guarantee efficient manoeuvre teaching. The second one is that HORUS allows to learn needle insertion, but since this is a complex manoeuvre, not all students can learn how to do it in only 12 h. The third one: HORUS not only allows to learn US-guided procedures, but also to select in 20 h students who will be able to realize that manoeuvre with no danger for patients. It is obviously difficult to assess the truth of those hypotheses, the last possibility is however seducing and would match the current exploitation of flight simulators in selective mode. In order to check those hypotheses, more experiences in simulator use are required. Having an evaluation of simulator feedback in real hospital practice of students would further be greatly valuable, which is currently not the case. We do however think that the third hypothesis seems very likely and that the simulator in companioning mode probably allows for the extraction of sufficient information on learning and realisation aptitudes of a complex surgical manoeuvre, so as to be characteristic of the realisation ability. In the end, we drew an obvious rule from those training experiences.

Rule 4: in companioning mode, a simulator can be used to evaluate learning abilities of a taught manoeuvre, method or reasoning.

3.1.3 Conclusions on Learning with the HORUS Simulator

HORUS has been the first simulator that we have developed and exploited for the training of our students. It proves to be very efficient for the training of US controlled exploration. We have extracted four generic rules from our HORUS experience. Food for thought further has to be deepened. Thus, with HORUS it turned out that teaching with companioning is more efficient than autonomous teaching. In fact, we did not turn these two facts into rules since one of the major drawbacks of companioning is the fact that it is teacher-dependent. Thus, if the teacher is not efficient, the simulator may not be exploited properly. No guarantee is therefore given through companioning. Conversely, our entire study on the HORUS simulator and the progressive improvement of its exploitation showed us that what can really guarantee learning success with a simulator is the development and respect of an educational approach using the simulator (Rule 2). When there is no automatic educational system, the teacher has an obvious advantage, since she/he is by nature pedagogy-oriented.

What would happen if we develop software integrating an automatic educational approach, just like video games integrating an increasing difficulty level in missions, so as to succeed in the more difficult missions through training on missions with increasing difficulty? In other words, can simulation offer efficient autonomous training? Is it possible to replace the advice on manoeuvres given by the teacher by an automatic system that guides these manoeuvres? When looking at the vocational education using simulators described and commented in this book, the answer seems to be negative since all the presented simulators are used in companioning mode. Conversely, when looking at the currently commercialized laparoscopic surgery simulators, the answer seems to be positive since they rely on the concept of autonomous simulation. In the case of the simulation of US-guided procedures, although we have today still not found an efficient method to replace the teacher throughout the entire instruction, we have been able to reduce her/his intervention down to a total of 5 instruction hours out of a total of 20 training hours, followed by a one hour exam. This entire procedure presents a highly satisfying end-result. It is thus sure that autonomy is possible if the simulator is used within the framework of a well-defined and controlled educational approach.

3.2   Learning Basic Manoeuvres of Laparoscopic Surgery

The use of ULIS during the non-residential traineeship during medical studies is fairly recent and only started in December 2007. The analysis that will be presented thus relies on a limited number of instruction sessions. It however allows to come up with some useful remarks within the framework of this book. Unlike HORUS, this simulator has not been evaluated on several teaching modes, but only in autonomous mode, since this was our initial objective. In this respect, ULIS fits into the same training format as the other currently commercialized simulators.

3.2.1   Autonomous Learning on ULIS

- **Familiarization with the simulator by students: simulator presentation**

As for the HORUS simulator, familiarization starts with a presentation phase of its functionalities and how it works. This presentation is done in the "demonstration" mode of the simulator, showing all the possible exercises but without recording the automatic evaluations provided at the end of each exercise. This initial companioning-type step is limited to 30 min, after which the teacher creates an account for the students so as to start their instruction. Since ULIS has been developed after HORUS, it features an automatic evaluation mode offering students fast and clear overview of their progress during the instruction. This "learning curve" is personalized and specifies a name, hence the necessity of creating an account. Furthermore, unlike HORUS, ULIS integrates an educative approach developed as progressive exercises. Rules 1 (necessity of automatic evaluation) and 2 (necessity of educational approach), which we have previously been defined, are thus respected. Rule 3 indicates the necessity of automatic and immediate feedback on the efficiency of a manoeuvre and its consequences. ULIS applies this rule for each of its exercises but in a voluntarily limited way. Thus, late camera displacement will make the target disappear and will in the end lead to a bad marking. Contact between camera and cavity will however only trigger force feedback on the tool (feeling of contact) and the deformation of the touched wall. Blur simulation of the laparoscopic view is thus not included. Liquid indeed blurs the optic when in contact with mucosae or a possible blood injury linked to the perforation of the cavity touched by the camera. These two last simulations would be a comprehensive and realistic illustration of the consequences of a bad manoeuvre. In the current version, the only consequence that is passed on from these errors will thus be a poor marking at the end of the exercise.

This is an educational choice and can be explained by the voluntarily increasing difficulty of exercises. Indeed, we do not want students to be immediately confronted with haemorrhage right from the first exercise. This issue is tackled in Exercises 5 and 6. Adding such options is technically possible and could be done in a second exercise session, confronting students with increased surgical realism, exposing them to their possible surgical errors. Rule 3 would therefore be perfectly matched. From the first session on, the consequences of these additions would logically be the death of the patient in case of severe haemorrhage and the exercise stop, moving on to a more complex surgical problem, without even being able to previously master the on-going exercise. This last point is of course not optimal when the objective is progressive and selective learning, which, unlike real simulation, is possible thanks to virtual simulation. Indeed, during surgical training or instruction on a living animal, students should compensate the previously described error by stopping the haemorrhage. From initial camera manipulation training, students would abruptly switch to the simultaneous manipulation and coordination of several tools, although they do not master the fundamental and mandatory tool manipulation enabling them to see and therefore to operate. By limiting this effect, virtual simulation allows to guarantee that the optimized educational approach is followed and respected.

We also have to point out that the virtual death of a patient, possible consequence of rule 3 if it is applied to the letter, raises other questions on potential dangers of a simulator on the psychological level varying from patient death becoming commonplace to possible trauma of students (though hardly probable). We therefore propose to clarify rule 3 with rule 5.

Rule 5: in simulation, automatic and immediate feedback on manoeuvre efficiency and its consequences has to be used according to the targeted educational objectives.

This rule can feed issues tackled in aeronautics regarding the virtual plane crash issue: has it got to go until the end of the crash as in real life or has it got to be stopped by the virtual simulator. In fact, we want to highlight the benefit and educational limits of simulation pushed to the ultimate situation. We think that if a simulator has to be able to reproduce the real surgical environment in a comprehensive manner, it also has to be able to limit that environment or to improve it thanks to virtual elements. This is one of the major benefits of a virtual simulator. The educational benefit of comprehensive simulation, which can go as far as the virtual death of a patient, is of course to show the consequences of errors that have been made. Students however will not necessarily know how to correct an error if they have not learnt it before. They are lost and the benefit induced by the realism of simulation is then largely questionable. Simulation is thus double-edged.

Fig. 9 Examples of exercises of the ULIS simulator: manipulation of surgical instruments (*left*), coordination between camera and an instrument (*centre*) and coordination between two instruments (*right*)

We therefore think that automatic and immediate feedback regarding the efficiency of the realized manoeuvre and its consequences has to be provided progressively and in an educational fashion. Feedback can be in text format during an initial learning phase, for instance indicating students that their manoeuvre would have led to a haemorrhage. Later it could be simulated realistically during an advanced learning phase, for instance simulating bleeding due to an error.

- **Use of the simulator by students**

At the end of the familiarization phase, students can train on their own account. They will then perform a set of exercises in order to learn and master respectively camera manipulation, surgical tool manipulation, camera and surgical instrument coordination, coordination of two surgical instruments, suction and coagulation (see Figs. 5 and 9). Each exercise has options making manoeuvres progressively more complex. When students reach a sufficiently high marking, they can move on to the next exercise. All marks are saved and the teacher can regularly check the progress of students without having to be physically present during the simulation sessions.

This phase has been realized over various durations between 10 and 30 h. After 10 h of use, students are not able to use all available options and their training is incomplete. After 20 h, students are able to do the exercises with normal camera configuration, but are unable to use 30° and 45° cameras.[11] After 30 h, students can do all the exercises, the results of automatic evaluation show however inequalities between students.

During the exercises, we have noticed that some students wished to be assisted or controlled by the teacher. In fact, the freedom of use entails the fear of not doing or learning manoeuvres properly. Although the software gives an automatic mark for each performed exercise, it cannot replace the reassuring presence of a teacher. Surgical training relying currently on companioning, this behaviour can seem natural, but it also points out one of the limits of automatic training. Furthermore, we have noticed that the initial use of the simulator systematically led to true enthusiasm in students whereas a prolonged 30 h use was at times wearisome for

---

[11]Cameras providing a view with a 30° or 45° angle with respect to the axis of the objective.

some students. Repetitive and isolated learning of surgical manoeuvres can explain this. The risk of this lassitude is that manoeuvres may become commonplace, losing their realism, and students may lose their motivation for that kind of learning. In both cases, the intervention of a teacher during the training cycle can be a solution to both reassure students and keep up their initial motivation. We have noticed that this intervention can be reduced to the review of student evaluation and to the execution of a very limited number of exercises under teacher supervision. Thus, doing several 10 h training cycles, including student evaluation at the end of each cycle, is a satisfactory solution to guarantee motivation and confidence. In the end, it is preferable to do three 10 h cycles rather than a single 30 h training ending with an evaluation.

Let's finally point out that we have used the simulator in two different environments: an IT room and one of IRCAD's training OR rooms. In the first case, students did not need to "dress up", i.e. wear the required compulsory OR outfit. In the second case, students not only had to dress up but were also confronted with a real surgical environment. We noticed in this second case a different approach to the simulator; manoeuvres never become commonplace and especially exercises become more realistic. Students were clearly more concentrated and trusted the simulator more, never comparing it to a game. However, we have to relativize these facts by the little number of students. A long term study will help checking this fact. This would mean that for learning even basic manoeuvres, real surgical conditions improve training.

- **Student evaluation**

At the end of autonomous simulator use by students, we make an "internal validation", which consists in checking on the simulator that a trained task has been acquired. This evaluation should ideally be done on an animal (real simulation), but because of the same economic or ethical reasons this is not possible. Unlike HORUS, we evaluated students on the same exercises as those that have been taught, favouring the most complex exercises. After 10 h, the chosen exercise will be carried out with a 0° camera. After 20 or 30 h, the exercise will be done with a 30° camera. In all cases, electrocoagulation or blood suction exercises due to haemorrhage will be accepted. Indeed, they have the advantage of requiring the simultaneous ability of controlling the camera, manipulating an instrument and coordinating camera/instrument. The teacher can then also choose to evaluate students on their weak points that have been revealed by the tracking of the automatic marking. The learning curve, which reproduces the marks saved at the end of each exercise, is indeed an efficient tool to spot the points students will have to work on.

After the evaluation phase, we have noticed that 30 h on the simulator seem to be sufficient to learn the taught basic manoeuvres, the best results being obtained by students who underwent three 10 h cycles. These evaluations show however disparities. Some students for instance are rapidly able to master manipulation after 10 h, while others have similar or weaker evaluations after 30 h. This could mean that the simulator would allow to spot students with more difficulties in performing surgical manoeuvres and who therefore require more training time. It will be essential to check that hypothesis during the rest of the standard clinical training. The simulator is today too recent to guarantee this hypothesis.

3.2.2   Conclusions on Learning with the ULIS Simulator

ULIS is the latest simulator that we have developed and exploited for training our students. It turns out to be very efficient to train basic manoeuvres in laparoscopic surgery. Its use showed us that autonomous learning is possible for the first phase of training, in particular thanks to the respect of rules 1–3. Autonomous training however has limits induced by the absence of a teacher. It therefore seems important to us not to leave students alone for too long. This has the advantage of exploiting the continuous evaluation so as to spot the potential problems of students early and to orient the training accordingly. In the current state of our experiences and knowledge, there is nothing more we can say, but those first very encouraging results for sure will allow us to pursue the approach of optimized training, in order to minimize the intervention of surgeon teachers because of their little available time.

4   Conclusion

In this chapter we have seen the exploitation of two training simulators for interventional procedures in radiology and minimally invasive surgery. They allowed us to define 5 simulator exploitation rules, in particular in autonomous mode. Our studies clearly show the feasibility of such training at the first training step. However, since we could not evaluate steps 2 and 3 of the training, there is nothing we can conclude for those two steps. The experience of other vocational training sessions using simulation will thus be very useful so that we can carry out those steps optimally.

Our work further highlighted an external consequence which was initially not expected. Indeed, surgical simulation provides a new vision of tomorrow's medicine. It improves student consideration of medical orientations which have difficulties in hiring the young generation, such as radiology or surgery. Attractiveness of new technologies and educational benefit of simulators give non-resident students the opportunity of carrying out manoeuvres which are usually forbidden thanks to virtual reality. This represents an attractive element for those difficult professions. Shall our hypotheses that we described in this chapter prove to be true, simulators could also allow to detect the skills and difficulties of each student earlier. This would ease recruitment and training of students according to criteria closer to professional abilities, essential point to guarantee tomorrow's medicine.

References

Amirian, M. J., Lindner, S. M., Trabulsi, E. J., & Lallas, C. D. (2014). Surgical suturing training with virtual reality simulation versus dry lab practice: An evaluation of performance improvement, content, and face validity. *Journal of Robotic Surgery, 8*(4), 329–335.

Forest, C., Comas, O., Vaysiere, C., Soler, L., & Marescaux, J. (2007). Ultrasound and needle insertion simulators built on real patient-based data. *Studies in Health Technology and Informatics, 125*, 136–139.

Hostettler, A., Nicolau, S., Soler, L., & Marescaux, J. (2013). Process and system for simulation or digital synthesis of sonographic images. *Journal of the Acoustic Society of America, 133*(2), 1201.

Nicolau, S. A., Vemuri, A., Wu, H. S., Huang, M. H., Ho, Y., Charnoz, A., et al. (2011). A cost effective simulator for education of ultrasound image interpretation and probe manipulation. *Studies in Health Technology and Informatics, 163*, 403–407.

Shaharan, S., & Neary, P. (2014). Evaluation of surgical training in the era of simulation. *World Journal of Gastrointestinal Endoscopy, 6*(9), 436–447.

Soler, L., Forest, C., Nicolau, S., Vayssiere, C., Wattiez, A., & Marescaux, J. (2007). Computer-assisted operative procedure: From preoperative planning to simulation. *European Clinics Obstetric and Gynaecology, 2*, 201–208.

Soler, L., & Marescaux, J. (2008). Patient-specific surgical simulation. *World Journal of Surgery, 32*, 208–212.

Tjiam, I. M., Berkers, C. H., Schout, B. M., Brinkman, W. M., Witjes, J. A., Scherpbier, A. J., et al. (2014). Evaluation of the educational value of a virtual reality TURP simulator according to a curriculum-based approach. *Simulation in Healthcare, 9*(5), 288–294.

Simulation to Train Hot Rolling Mill Operators in Metalwork

Gérard Bonavia

Metal in hot rolling process

How training and transfer of professional knowledge can be used to accompany the generation staff renewal, to improve industrial performance and create new working relationships?

In a room in the Rolling Mill Department at ArcelorMittal's industrial plant in Fos-sur-Mer (France), four groups of two operators of all ages, prepare to roll metal on computer workstations. They discuss amongst themselves with the help of two trainers. After this sequence, they gather around a table to talk again of what they have done, the results obtained and what guided them in their decisions.

G. Bonavia (✉)
Private Consultant in Engineering of Steel Industry, PACA district, France
e-mail: g.bonavia@free.fr

© Springer International Publishing Switzerland 2016   119
Ph. Fauquet-Alekhine and N. Pehuet (eds.), *Simulation Training:
Fundamentals and Applications*, DOI 10.1007/978-3-319-19914-6_6

Discussions on what they experienced, justification of their choice... They put forward their arguments, listen, challenge each other, ... Welcome to the steel industry and an operator training session on a hot metal rolling mill simulator.

Exchanges of experiences, comparison of points of view, virtual exercises, ... We are a long way from an academic lecture.

1 Metal Rolling Mill Operator: A Changing Profession

Insert 1 Hot train rolling mill

Hot rolling is a step in the steel process that aims to give the dimensions and the appearance desired by the client. It is designed to reduce the thickness of the slabs of steel (parallelepipeds of steel 20 tons on average and 225 mm in thickness, between 6 and 18 m in length and from of 600 to 2200 mm in width) using 7 cages on the train band.

The 7 cages are spaced 6 m apart and each is equipped with 4 cylinders. The product is processed in the 7 cages at the same time. The electric power of each cage is 9.6 megawatts. The reduction in size is obtained through crashing and being stretched between cylinders to obtain a final thickness between 1.5 and 22 mm with an accuracy of approximately 10 mm. The strip of metal can be up to one km in length. The mill is automated, with two operators controlling and regulating the process. They also intervene during the periods of inactivity to service the mill with the help of electricians and mechanics.

Despite the progress of automation on the production lines in the steel industry, stability and performance of our industrial results depend on our experienced operators. The quality of the decisions and actions they make in real time, based on changes in the process, the state of the installation, the variety of manufactured products, participate towards the results of our production.

The profession of metal rolling mill operator (hereinafter RM operator) is not at its first change. With the advent of mechanization, more and more sophisticated tools have replaced the physical strength of production personnel. The first automation moved the operator from the tool to centralized remote cabins. These automations were an advantage that experienced operators had to do without. The performance and reliability of computer systems justified this attitude. Today computer systems have progressed, they have become essential. Their downtime often means a break in production.

Some might be tempted to see the RM operator as a supervisor of automation. It would be a serious mistake and a misleading for our results. The RM operator is the regulator of the production system. S/he overrides faulty automation. S/he can cope with unexpected situations and formulate solutions not listed in the procedures. The action or non-action are inseparable from the decisions s/he takes in anticipation or in response to events faced. The training should take account of these developments as well as evaluation criteria. Previously, they were focused on learning the professional gestures, they are now developing comprehensive understandings of the production system. Orientation in the entire process, the cooperation established with those working upstream or downstream and the flow of information are at the heart of the activity of the RM operator.

2 Training for Rolling Mill Operators

Massive departures of RM operators from the "baby boom" generation make training of production operators a major challenge for the plants of ArcelorMittal group.

Insert 2 ArcelorMittal—presentation

ArcelorMittal is the world leader in the steel industry, with 310,000 employees in more than 60 countries. The company merged the first and the second largest steel producers in the world, Arcelor and Mittal Steel. ArcelorMittal is the leader in all major global markets, including automotive industry, construction, household appliances and packaging.

The company is a leading actor in the field of technology and R&D and has significant resources of raw materials as well as excellent distribution networks. Its industrial facilities are spread over 28 countries in Europe, Asia, Africa and America and permits the company to be present in all key steel markets, both in emerging and developed economies. The company is developing its positions in China and India, two countries whose markets are booming.

On the rolling train of Fos-sur-Mer (Fig. 1) we responded to this challenge by designing and integrating a simulation training in the professionalization curriculum. The work on the simulated production line is designed for two operators (Fig. 2). They must constantly work in cooperation. The distribution of activities between the two operators is not presently optimized when technological developments and the production needs are taken into account. A reflection for new ergonomic design of the control panels is in progress. It is mainly based on the analysis of activities carried out on simulator. These ergonomic considerations have been validated by the operators during the first training sessions.

Fig. 1 Hot rolling train in Fos-sur-Mer (France)

Fig. 2 Operators in control room in real operating situation

There is no academic training for these professions. So far, this training has been based on two pillars: the contribution of theoretical knowledge and peer companionship.

The former is a contribution by experts from technical departments developed around the processes main themes. This knowledge is crucial. It relies on an experienced technical and scientific basis with a precise lexis but it also has shortcomings. Indeed training is academic, designed by people familiar with this mode of transmitting knowledge. However it is not always suitable for operators who are not comfortable with handling theoretical concepts. It dissects the process into sequences, thus permitting the expert to go into the detail of each theme.

This does not reflect the work carried out by the RM operators who must constantly be aware of the situation in its globality, sum up all the information and take decisions in real time. The appropriate regulation of the process assumes choosing the right compromise every time by identifying the actual causes of dysfunction and deviation in a global context. For all these reasons, an operator expecting for answers to his/her problems might find this kind of training disappointing. Memorization of the information given is poor.

The other axis of training is companionship of the novices on the workstation by more experienced co-workers. This axis is beneficial because it creates social links in teams between experienced workers and novices, between old and young operators. Unfortunately this mode of transmission also spreads myths, old "recipes" with a approximate vocabulary. It leads to confusion between the causes and the symptoms with little reflection on the meaning. Young operators are consequently limited to certain team practices.

3 A Simulator for Training

In 2002 a multidisciplinary project team designed and built the training simulator for the band trains of rolling mills. This team consisted of RM operators, computer scientists and a specialist in professional didactics.

3.1 Why a Simulator?

Without abandoning the aforementioned two axes, simulator training was expected to compensate for the disadvantages induced by the lack of practical and operational aspects of academic training. It should also avoid the approximate and limiting nature of companionship. In addition to the indispensable deepening of basic phenomena, situations that do not always have a unique explicit solution must also be addressed.

Training on simulator is a pragmatic response to the need for skills development. By means of interactivity, it permits the development the operating skills through action. The scenario make it possible to address concrete issues encountered in real operational situations. Discussions on the logic of actions, justifications for decisions encourage the pooling of scattered information and the collective resolution of problems. It is an opportunity to structure the profession of RM operator. It allows a conceptual approach on the basis of observed and therefore concrete facts. This is a time for discussions between individuals from the same profession and for sharing the words of the profession.

3.2 Which Kind of Simulator?

When the idea of training with a simulator sprouted, several views were expressed. The engineers saw it as a technical challenge: building a virtual rolling train as a clone of the "real" one. The computer engineers saw a way to test and develop software and models. It also rose concerns amongst operators: a new computer was coming; after automation, after elaboration of procedures and streamlining work through technical and safety standards, they saw it as yet another way of having their activities controlled. When thinking about training we also imagined a simulator close to the reality in the field. Operators could go and train on the simulator during breaks. It would be available as self-service.

What we have today is very different from what everyone had imagined. A professional didactics specialist helped us and guided us in our choices. She left us enough freedom for this simulator to be ours. Retrospectively we agree that we could have listened to her a little more and not go as far in our desire to copy reality. The simulator is inseparable from the related pedagogical approach. In fact, it is not a self-service simulator. It is far enough away from the workplace for operators who are trained here to not be disrupted by the operation of the real production line. All courses are undertaken in the presence of two trainers.

> **Insert 3 Trainers**
>
> The trainers work in pairs. One is a technical department representative. S/He is not a RM operator but has solid technical and scientific knowledge that enables him/her to verify the information submitted and to supplement it if necessary at the appropriate time. He/She is trained in management and pedagogy for group training. The other trainer is a peer of the trainees. He/She is qualified to talk about practice of rolling. He/She is recognized by peers.

In contrast to flight or nuclear reactor simulators mentioned in the present book, we have developed a simulator which only partly reproduces reality. The principle remains that of interactivity: the simulator must respond to solicitations in real time and the RM operator must recognize the responses. To build this type of simulator it was important to answer three questions:

- What part of reality can be left out? Any part that is not present will be mentally rebuilt by the operator during the training session. The reconstruction of this missing environment is discussed during the training session debriefing.
- What must be shown with a high degree of fidelity? Experience shows that an even imperfect model of physical phenomena is sufficient. Some simplifications have been made (linearization, statistical modeling, etc.). Yet some elements require a realistic representation so that the trainee recognizes the job. The most important are, on one hand, the interaction between different physical phenomena of the process, and, on the other hand, the feedback of the temporal dimension.

Fig. 3 Training simulator

- What should be shown to operators as it cannot be seen or remains unclear in their real working environment? The view of certain phenomena is improved for educational purposes.

Making these choices presupposes a good knowledge of the profession of RM operators and of their social environment. Before studying the design of the conception study of the simulator, the activities are analyzed. "The study of the profession by those who work in it, less than by those who design it" as defined by our professional didactics specialist Dr. Stéphanie Guibert (Ph.D. in Science of Education).

To direct us within these guidelines (analysis of the activities and teaching methods), in addition to ICT[1] technicians and RM operators the design team's skills must include work psychologist, pedagogist, didactics and cognitive work analyst.

This simulator, one of the first in the steel industry, is now located on other roll mill plants of the company in Dunkirk (France) and in Spain. Despite cultural and equipment differences, as well as a different rolling direction (operators are positioned on the opposite side of the rolling mill train), Spanish RM operators have no difficulty using the simulator. This proves that the representation and modeling of rolling we developed are sufficiently generic. This also proves that what structures and organizes the activity of the operators is not a control panel but the fundamentals of the profession of a RM operator (making a diagnosis, reasoning, making a decision, controlling the actions, working in cooperation).

The simulator (Fig. 3) is made up of a central screen that represents the rolling train, the product being rolled and key indicators which evolve in real time.

---

[1]ICT: Information and Communication Technologies.

The screens on the sides give information about input and incoming material on one hand, and indications about output on the other hand. A touch-screen on the table provides functionalities of the control panel and allows the operator to act in advance and during the ongoing process.

4 Organization of the Training Process

All operators (40 people) undergo training on the simulator every two months by groups of eight. Trainers prepare training, teach and assess the effectiveness of the simulation training.

Trainers are not specialists in pedagogy. Some basic rules may nevertheless help them:

- Training must clearly indicate the intended operational objective and expected results.
- For the exchange between participants to be effective, it must be based on a considerable work beforehand regarding preparation, formalization, organization and storyboarding. This preparation takes material and pedagogical aspects into account.
- Do not forget that training is a moment of discussions where trainers themselves receive relevant information for their own job.
- The person must act and demonstrate before explaining.
- There is a need for the trainers to put themselves in the place of the trainees. Make sure that concepts and definitions used are understood by everyone. Field words must be used.
- Trained operators seek information which they can assimilate, based on their own centers of interest and which can help them work.
- Manage participants by keeping the lead: respond to comments and questions, accept discussions, and note the questions to which we cannot respond immediately... Always go back to the subject being taught.
- Later on, the transfer of what has been learned must be monitored and the implementation carried out in the field must be assessed to verify the efficiency of training.

For reasons of efficiency and coherence between trainers, the following three sequences are standardized: before, during and after training. Tables 1, 2, 3 and 4 are in all the training files. They give procedures for each stage.

4.1 Prior to Training

The tables describe in detail the steps in the overall training process with a simulator. They specify what to do as any appropriate standard does: how to do it. They

Table 1 The stages that precede the training session

Stage	What to do	How to do it	The persons responsible
Planning the training	Define the topic Define the stakes Designate the trainees and trainers Set the month of training	With the help of the training plan: differences between target team and real team	Training Committee
Organization prior to the training session	Set the date, the duration Designate trainers Identify participants Prepare practical aspects/logistics Request an attendance code Inform the workshop and the line manager	With the help of the training table Book room, hardware, convocations, timesheets, meal tickets, ...	Work station manager Trainers, human resources office
Preparation of training session	Define operational objectives Develop the exercises and the scenario for the session Choose lecturers/trainers if appropriate Identify real cases that are problematic Link training with safety concerns and procedures	With the help of performance objectives and field issues Timing, rhythm, preparing associated media, think of key points for each sequence Help trainees participate, Identify those who will assist in transferring knowledge Prepare safety instructions, procedures and standards existing in the concerned field	Trainer Trainer and tutor's area Trainer

Table 2 Training sequences

Stage	What to do	How to do it	The persons responsible
Training session	Introduction: presentation of participants if necessary Why are we together? Start-up Alternation between simulator and debriefing Make trainees adhere to the objective of the training session Encourage trainees to participate	Reminder of the rules: we are together to learn Training exercise Approximate ratio: 20 min on simulator, 1 h of debriefing Establish link with performance Encourage trainees to ask questions to ensure understanding	Trainer(s)

Table 3 Table for verification that the training session addresses all aspects of the situation

Stakes	External: needs, customer's requirements Internal: quality (downgraded, repaired), reliability
Causes	Product: format, nuance, client, length, weight, … Process: templates, modifications, regulations, offset, manual Tools: condition, cylinders, fluid, …
Operator's actions	HMI[a]: indicators, actuators, panels, screens Actions: planning, anticipation, reaction
Context	Located on the product Isolated case, series Flow effect, cadence Degraded steps
Linked activities	Visits Calibrations, testing Tool changes

[a]HMI: Human-Machine Interface

Table 4 End of the session

Stage	What to do	How to do it	The persons responsible
At the end of the session	Build a strategy for implementation in the field with measurable performance indicators	Give everyone a "roadmap": What one should remember What one should you do Tracking indicators	Trainer
After training	Follow-up on selected indicators Individual follow-up of acquired knowledge Technical support after training	Inform the workshop and line managers of the indicators monitored Compliance with the output sheet Access to documents, phone number for technical support	Area operation manager Tutor Technical dept

help trainers so that they do not forget anything essential and succeed the training session.

The educational objective is defined from training objectives: skills to develop, enhance and maintain. What precise results must the professional attain at the end of the training program?

In addition to the physical and logistical aspects (convocations, reservations, etc.) the preparation has four specific objectives:

- Sharing of the objectives of the training sessions and their issues between all involved people (managers, trainers, human resources department, and technical departments). When possible, choose or develop indicators that will measure the effectiveness of the training.

- Establish what the output document will be at the end of session. It should be 80 % ready. Trainers must accept surprising ideas that might come up from the trainees.
- Involve trained operators by asking them to express their expectations on the topics of the training sessions.
- Prepare the training session itself: gather documentation, develop exercises (Insert 4), and organize pedagogy according to the group of trainees (homogeneous or heterogeneous). Simulator exercises are designed for staging of situations experienced in the field related to the chosen educational objective. Learning in a simulated situation is oriented towards implementation in the field.

Insert 4 Elaborating training exercises

Specifications of the scenario: number of products, information on the state of the line, event in the cabin, setting priority of objectives, planning, incoming material...

Distracting event: faulty product, error, computer problem, tool malfunction...

Printing of the program: information available about the program

Indicators displayed on input screen: product, state of regulatory functions.

Indicators displayed on central screen: visualization of the product, of the installation, forces, positions, alarms...

Indicators displayed on output screen: graphs of results

Touch screen (panel): actuators at disposal for anticipation and rolling.

4.2 The Training Session

The simulated situation incorporates computer hardware (the simulator) and develops discussions mixing the following:

- practice and technology,
- logic of the installation and its users,
- experience and science.

The session is organized in order to:

- develop an operational understanding of hot rolling
- share a common repository, a representation of the activity,
- give meaning to one's assignments, one's function,
- identify organizing systems of the activity of reasoning.

The table below shows the pace of the training session. It details the objectives of the different sequences and the means to achieve this.

The session begins with an introduction. This is the opportunity to define the rules (Insert 5). An easy exercise helps to get going on the simulator. It addresses missing elements and the mental representation of the situation.

Insert 5 Start of the simulation

The session begins with the statement of principles and rules:

- no notion of passing an exam
- no traces stored
- trainees are rolling mill professionals

Objective: to learn

Presentation of the simulator.
An exercise is undertaken to familiarize trainees with the simulator.

The training session, after the introduction, continues with alternating sequences. Each sequence consists of an exercise with the simulator and a collective debriefing session.

There are 8 operators in each group and are divided into four pairs on the workstations. Everyone does the same exercise at the same time. The different sequences are of increasing difficulty. It comes from the fact that it mobilizes increasingly reasoning and knowledge to identify the disruptive element.

The first exercise (Insert 6) aims at sharing technical language, the words of the trade.

Insert 6 First simple exercise

Objective: language, technic, aspect ratios operator/engine, indicators, actuators.
It is an easy exercise:
The draft is perfect.
Cages 5, 6 and 7 are unstable.
When the operators see unsuitable aspect ratios, they are corrected, and thus applied to following products.
This is a moment when language begins to be shared and provides clarification.
A question must be asked: what is a faulty aspect ratios? What are the possible causes? What other indicators provide information? Define technical words or expressions at the appropriate time.

The second exercise's objective (Insert 7) is the need to understand the links between two phenomena. It is a reflection between the symptoms (what is apparent) and causes.

Insert 7 Second exercise about interrelations between phenomena

Objective: Working the link reduction-symmetry and concepts of C/H
The draft is perfect.
Aspect ratios are seen everywhere and yet only cages 1, 2, 3 and 4 are not as expected.
The solution comes from reasoning and corrections: beforehand or further down the line.

The third exercise (Insert 8) does not have any satisfactory solution using the actuators available to operator. Its resolution will come from cooperation established between colleagues beforehand.

Insert 8 Third exercise: cooperation

Objective: to establish cooperation.
Verbalize: "There is a defect and it is a colleague who has the actuator to correct it".
The initial state of the facilities is the same as for the second exercise but the draft has a defect.
There is never satisfactory solution without reducing in size. Aspect ratios must be chosen. Only one solution exists, RM operator must ask his/her colleague operating the reducing machine to correct the draft.
The RM operator learns to solve problems through cooperation.

4.2.1 Training on Simulator

At the beginning of each exercise, trainers provide trained operators with the rolling program they will have to apply. As in the real control room, some have a glance at the program then start rolling, others take the time to study it, sometimes identify anomalies and inconsistencies. The training session has now really begun. Soon differences appear between good RM operators, those who analyze and anticipate, and those who often meet with failure.

While trainees "roll", they face problems similar to those they encounter in real operating situations. After a diagnosis of the state of the situation, they develop strategies, decide and act on their control panel and can control the results of their choice on indicators. The eight operators in the training session are split into groups of two so as to discuss and develop a binomial decision. The trainers answer any questions, provide information during the exercise. They check whether the communication between trainees is good or not.

4.2.2 Debriefing of the Simulator Run

When the exercise ends, training operators and trainers gather around a table. Debriefing of simulated situations starts with actions made on each workstation.

The main question of the discussion is: "for what reason this choice, this decision, and this action?"

- which indicators were used when reasoning? Which indicator can help them diagnose present or future problems? What permits them to validate or invalidate a choice of action or non-action?
- What actuators to use? How are they handled? In which order?
- What other phenomena are related to the problem?

At this time the information discussed is validated and knowledge is shared collectively. The logics of actions carried out are compared and the trainees explain their actions and decisions. A generalization is developed from the case study regarding the different categories of situations. At all times trainers ensure that the transmission of information from one speaker is adapted to the listener's references and not only that of the speaker. Meaning must be provided; it is better to make trainees speak rather than let trainers lecture.

Skills development is carried out through oral presentation of one's own practices, justified reasoning and through listening to others.

Trainers can also identify potential improvements for the organization and ergonomics of control panels.

About 1 h of debriefing is necessary for about 15–20 min of simulator.

4.2.3 Situations Experienced in Training

Despite precautions taken including standards of preparation and having qualified trainers, certain situations may cause difficulties for trainers:

Disagreement between the experts. The chosen educational strategy depends on of the constitution of the group attending the training session. Is it a homogeneous group? Are they beginners or experts? Is it a group of mixed ages and levels? With a heterogeneous group, trainers rely on experienced operators. The solution to the problems that trained operators expect does not come from the "teacher" but from co-workers. The trainer becomes the one who connects people, who encourages this discussion that bonds the group and validates the relevance of the information discussed. The trainer intervenes, if necessary, to provide further technical and scientific as well as to make it applicable in other situations. With this educational strategy, the danger is to have two experts in the group who do not share the same point of view. The management of the group can become complex, the referent RM operator's help is crucial. We must prepare a priori consensual solutions.

Transgression of rules by experts. The operating procedures and standards mark out certain practices. Each voluntary deviation must be recorded. How to behave when an expert (seen by others as such) offers border-line training solutions, and sometimes goes beyond the limits authorized? Intransigence is de rigueur on safety issues. The law is the same for everyone whatever their level or skills. Under no circumstances unsafe practices should be validated. In certain areas, it might be interesting to underline some "underground" practices through discussion, sometimes to deconstruct them or sometimes to officialize them. In his conclusion, Marc Labrucherie (chapter "aircraft") deals with "voluntary deviations", talks about the inescapable rules and about the part that should be left to the intelligence of the workers giving then acceptable leeway.

The session deviates towards non-simulated cases. The simulation offers selected simulated operating situations. The temptation is great for trainees to talk about the last event experienced in the field. It is important not to get sidetracked and to stick to the topic. Generalization and conceptualization are elaborated from the actual action on the simulator without forgetting the goal of the training session. The training materials help the trainers to remain centered on the topic of the session. Preparation will pull all information on the topic. Table 3 recalls the content that needs to be addressed.

The trainer is not credible. s/he does not know how to roll and has no technical background. The trainer is sometimes confronted with very experienced rolling professionals. S/He has never worked himself in the field. We have circumvented this difficulty by bringing in a co-trainer with a referent RM operator. They work together from the beginning in preparing and developing the exercises. The referent RM operator is the person who is most often responsible for verifying that the resolutions adopted are applicable and applied in the field.

Failure in simulation. The question of failure is much less problematic than in other parts of the present book. The death of a patient or the crash of a plane are not at stake with rolling. It is a question of incidents and not accident. The reliability suffers, operators' pride too, but nothing traumatic happens.

4.3 End of the Session and Afterwards

The timing of the training session should reserve the last half hour to share resolutions that will then be implement. The table below describes two milestones: what happens at the end of training and what will happen afterwards.

The actual outcome is a "roadmap" which includes what should be remembered from the training session. Whenever possible, a standard of implementation in the field or a strategy of actions for a category of situations are determined most of the time. An indicator will show that training is appropriate and that progress has been assessed. This field is the most relevant evaluation. The assessment is still a delicate but yet indispensable concern. Training is an investment (resources out of production, preparation time and time in training, cost) and the assessment participates towards measuring the return on investment?

Do we want to evaluate the trainee or the quality of the training program? Beware of "school" type assessments or multiple-choice tests (often seen in the company). We must never forget that the trainees when evaluated adapt themselves to the evaluation system. They render (regurgitate?) what the assessor wants to hear rather than what they understood. What does this type of evaluation provide? What does it certify? In addition to the choice of relevant follow-up indicators (compared before and after training), assessments are also carried out by the tutors at workstations. These evaluations allow us to verify that training decisions are applicable in work situations by the participants, they attest to the quality of the training session. We also check that each trainee can apply them: were they well assimilated and can they be implemented?

Operational skills you want to develop are those that explain the performance. The reference is not knowledge, know-how or personal conduct but the activity (action, communication, anticipation, collaboration...).

What is interesting at first and that must be assessed is not the achievement of the task but the relationship between the trainees and the task, their understanding of the purposes, and the orientation of the process, of the integration in the organization.

5 Training and Professional Practice

The improvement of the RM operators' professional practice does not depend on an accumulation of knowledge but on their appropriate use in situation. What characterizes efficient RM operators is their ability to act and to anticipate a problem.

Skills depend on processing of available the information that must be understood in order to analyze the situation and implement appropriate solutions. RM operators must use procedures to apply known solutions but must also elaborate solutions when facing rare situations.

The actual approach of operating implies a close collaboration between operating teams, maintenance teams and technical support teams. It requires adaptation to technical changes in the industrial installations, in automation developments and in organization.

Simulation training sessions based on reconstructed situations and collective problem solving are a relevant and pragmatic response to the improvement of professional practices for RM operators.

The simulator is hardware that permits RM operators to learn a range of actions related to the cases studied but in itself does not allow development of strategies adapted to all real operating situations. Training contribution is based on the trainers' work which federates the participants on problem solving, thanks to the simulated situations and the pedagogical strategy.

Training focuses specially on the development of skills related to the practice of collective operation in an industrial process whatever the age of workers and their level of experience may be. It allows the integration of newcomers into teams through shared elaboration of knowledge of the profession.

Conclusion

Philippe Fauquet-Alekhine and Nane Pehuet

All along this book, we have tried to propose ways and pragmatic advises in order to give the readers matter to think and make their own professional practice to evolve.

With the last chapters, we wish the readers to have questioned their own practice through concrete examples and under the light of theoretical contributions regarding pedagogical methods given in chapter "Reflections and Theoretical Contributions Regarding Trainers' Practice and Simulation".

The question of simulation training assessment still remains and is not easy to deal with as it is straight off a social problem. Perhaps it will be the subject of a future book.

It specifically takes all its importance when the return on investment must be rated, as Bonavia or Geeraerts and Trabold point out in their chapters addressing respectively steel rolling mills or anesthesia.

In fact, simulators are costly facilities. Companies which are involved in such financial investments must be able to rate whether this kind of action allows them to reach the goals, from a technical or logistical standpoint and for a possible future adaptation of the training program.

Some evaluation criteria may contribute to rate the financial feedback. The reporting of the company performance also contributes to this aim, as one of the training goals is to improve the performance related to different aspects such as security, safety, availability, time respect, and client satisfaction.

Ph. Fauquet-Alekhine (✉)
Human Factors Consultant, NPP of Chinon, BP80, 37420 Avoine, France
e-mail: philippe.fauquet-alekhine@edf.fr
URL: http://www.hayka-kultura.org

Ph. Fauquet-Alekhine
Laboratory for Research in Science of Energy, Montagret, France

N. Pehuet
INSHS/CNRS, Paris, France
e-mail: nane.pehuet@cnrs-dir.fr

© Springer International Publishing Switzerland 2016   137
Ph. Fauquet-Alekhine and N. Pehuet (eds.), *Simulation Training:*
Fundamentals and Applications, DOI 10.1007/978-3-319-19914-6_7

It will yet always be difficult to establish a trustworthy correlation between training and real operational performance, especially because results always depend on factors which are difficult to separate from one to another.

It must be underlined that these practices and pedagogical methods linked with the simulation training will be adjusted according to the technological progress which will give new complex socio-technical systems; the latter will require, at this time, new training methods to be invented. These developments should permit to plan more powerful simulators, both for virtual and non-virtual simulation.

The simulation training is a broad domain that remains to be discovered. So, our work is not over as it is part of a broader phenomenon in constant development. And this is rather interesting for us since everything remains to be done, in a better way!

The reader will find matter to think the topic by having a look at the following books which develop some of the themes discussed here:

The power to act:
Clos, Y. (2008). *Travail et pouvoir d'agir*, PUF.
The author finds that work makes women and men tackle social events in which the outcome will have a heavy weight on the fate of future generations. He raises the question whether work analysis will be able to assist individual and collective actions needed to cope with it. Y. Clot makes an inventory of the historical, theoretical, methodological and technical resources available in Work Psychology to develop the power to act of people in their professional environments.

The error:
Reason, J. (1990). *Human error*. Cambridge University Press.
The author develops the theory of latent errors, either in the structure of process, organization, management or improvement of skills.

The reliability:
Amalberti, R.; Mosneron-Dupin, F. (1997) *Facteurs humains et fiabilité: quelles démarches pratiques?* Octares Editions, 1997.

In a simple style, the book introduces the main concepts of reliability and addresses more the experience of intervention in work analysis, its conditions of success and its limits, than theories or methods that would renew approaches regarding safety of technical systems. This book seeks to provide the reader a working knowledge rarely mentioned by specialists but which is essential to the success of the intervention aiming at making reliable socio-technical systems.

Verma, A.K., Ajit, S., & Karanki, D.R. (2010). *Reliability and safety engineering*. London: Springer.
The authors present reliability terminology in various engineering fields and describe the latest applications in the area of probabilistic safety assessment, such as technical specification optimization, risk monitoring and risk informed in-service inspection. Reliability and safety studies must, inevitably, deal with uncertainty, so

the book includes uncertainty propagation methods: Monte Carlo simulation, fuzzy arithmetic, Dempster-Shafer theory and probability bounds. Reliability and Safety Engineering also highlights advances in system reliability and safety assessment including dynamic system modeling and uncertainty management. Case studies from typical nuclear power plants, as well as from structural, software, and electronic systems are also discussed.

Further Readings

Simulation and Training

Bainbridge, L., & Quintanilla, A. R. (Eds.). (1989). *Developing skills with information technology.* Chichester: Wyley & Sons.

Boet, S., Granry, J. C., & Savoldelli, G. (2013). *La simulation en santé: de la théorie à la pratique.* Berlin: Springer.

Fuchs, P., & Moreau, G. (Eds.). (2003). *Le traité de la réalité virtuelle.* Paris: Presses de l'école des mines de Paris.

Hays, R. T., & Singer, M. J. (1989). *Simulation fidelity in training system design: Bridging the gap between reality and training.* Berlin: Springer.

Pastré, P. (Ed.). (2005). *Apprendre par la simulation. De l'analyse du travail aux apprentissages professionnels.* Toulouse: Octarès.

Patrick, J. (1993). *Training: Research and practice.* London: Academic Press.

Risk, Safety and Human Error

Hollnagel, E. (2009). *The ETTO principle: Efficiency-thoroughness trade-off. Why things that go right sometimes go wrong.* Aldershot: Ashgate.

Hollnagel, E., Woods, D. D., & Leveson, N. (Eds.). (2006). *Resilience engineering. Concepts and precepts.* Hampshire: Ashgate.

Pham, H. (Ed.). (2011). *Safety and risk modeling and its applications.* Berlin: Springer.

Reason, J. (2008). *The human contribution: Unsafe acts, accidents and heroic recoveries.* Burlington: Ashgate Publishing Company.

Roeser, S., Hillerbrand, R., Sandin, P., & Peterson, M. (Eds.). (2011). *Handbook of risk theory: Epistemology, decision theory, ethics, and social implications of risk.* UK: Springer.

Appendix

Cross-thematic Table

Themes	Key words	Aircraft	Nuclear reactor	Anesthesia	Surgery	Rolling mills	Theoretical contributions
Activity transformation	Genre		79, 80				14–16
	Knowledge	45–47, 52, 53	62, 69, 76, 80			122, 124, 130, 132, 134, 135	7–9, 14–18, 20, 22, 25, 26
	Know-how	52, 55	60, 61, 68, 69, 80			134	7, 17, 20, 24
	Style		79				13–16
	Zone of proximal development						9, 22
Assessment	Assessment	47, 57	69, 77	93		134	20, 21
Briefing	Briefing	35, 45, 46	64, 65				8, 19
Companionship	Companionship					122, 123	
Competencies/ skills	Competencies/ skills	34–36, 44, 47–49, 52–55	60, 62, 68, 69, 80	90, 94	116	123, 125, 128, 132–135	2–9, 15, 18–26
Debriefing	Debriefing	35, 47–50	60, 63–68	89–93		124, 127, 130, 132	3, 8, 9, 11–24
Discussion (put into ...)	Controversy						16, 22
	Collective word		73				
	Metaphor						16
	Silence	50					16
	Speech		68, 74, 78, 83				16, 17
	Vocabulary					123	

(continued)

© Springer International Publishing Switzerland 2016

Ph. Fauquet-Alekhine and N. Pehuet (eds.), *Simulation Training: Fundamentals and Applications*, DOI 10.1007/978-3-319-19914-6

(continued)

Themes	Key words	Aircraft	Nuclear reactor	Anesthesia	Surgery	Rolling mills	Theoretical contributions
Distanciated posture	Distant	46	67–69, 73				9, 16, 17
	Distance						12, 17
Ethics	Ethics				107		8, 22
	Respect(ing)	50–53	61, 68				4
Leadership	Leadership	47–49	77	90, 94			8
Observation	Grid	48					11, 17
Performance	Failure	46, 52, 53				131, 133	7, 13, 17, 20, 25
	Performance	31, 34, 36, 43, 47–55		92, 94		120, 127, 128, 134	6, 10, 18, 22, 23
	Success			93	111		11, 19, 20
Reliability	Reliability		80			120, 128	17
Retrodiction	Retrodiction		77				13, 16
Risk	Risk	35–40, 42, 43, 52, 57	71, 77, 79, 81–84	87–89	95, 104, 106, 115		1, 6–12, 16, 17
Scenario	Class	39–43, 46–50					24
	Phase		66, 72, 81	88	104–107, 110, 112, 114, 115		3, 5, 11, 19, 21, 24
	Scenario	32, 35, 38	61, 65–75, 82	92, 93	97, 110	123, 127, 129	5–11, 18, 19, 26
Situation (limit)	accident/death	55		88, 93	106–113		
	Limit						6, 7
Simulator (type)	Artifact						2, 4, 19
	Autonomous				97, 103–116		10
	Full scale	32	60, 61, 65				3, 4, 28
	Full scope						3
	High fidelity			87, 90, 93			
	Partial						2, 3
	Realistic	40	61	87, 89	97–102, 113–115	124	
	Virtual				96, 97, 113	124	4
Trainers' practices	Taking notes	35, 47	67				11, 12
	Trainers' practices						25, 26

(continued)

(continued)

Themes	Key words	Aircraft	Nuclear reactor	Anesthesia	Surgery	Rolling mills	Theoretical contributions
Training (posture)	Enveloping		67, 69				9
	Inserted		67				9
Trainers' training	Trainers' training						24
Trust	Trust	46, 49, 54, 70	70				8, 13, 19, 23
Video	Video		60, 61, 73, 76	90	111		12, 17, 28
Voluntary deviation	Transgression			93		133	
	Deviation	54, 56				133	11

Index

© Springer International Publishing Switzerland 2016
Ph. Fauquet-Alekhine and N. Pehuet (eds.), *Simulation Training: Fundamentals and Applications*, DOI 10.1007/978-3-319-19914-6

Printed by Books on Demand, Germany